Finance is one of the fastest growing areas in the modern banking and corporate world. This, together with the sophistication of modern financial products, provides a rapidly growing impetus for new mathematical models and modern mathematical methods; the area is an expanding source for novel and relevant 'real world' mathematics.

In this book the authors describe the modeling of financial derivative products from an applied mathematician's viewpoint, from modeling through analysis to elementary computation. A unified approach to modeling derivative products as partial differential equations is presented, using numerical solutions where appropriate. Some mathematics is assumed, but clear explanations are provided for material beyond elementary calculus, probability, and algebra.

This volume will become the standard introduction for advanced undergraduate students to this exciting new field.

The Mathematics of Financial Derivatives

For our children
Oscar, Zachary and Toby

The Mathematics of Financial Derivatives

A Student Introduction

PAUL WILMOTT

University of Oxford and Imperial College, London

SAM HOWISON

University of Oxford

JEFF DEWYNNE

University of Southampton

PUBLISHED BY THE PRESS SYNDICATE OF THE UNIVERSITY OF CAMBRIDGE
The Pitt Building, Trumpington Street, Cambridge CB2 1RP, United Kingdom

CAMBRIDGE UNIVERSITY PRESS
The Edinburgh Building, Cambridge CB2 2RU, UK http: //www.cup.cam.ac.uk
40 West 20th Street, New York, NY 10011-4211, USA http: //www.cup.org
10 Stamford Road, Oakleigh, Melbourne 3166, Australia

First published 1995
Reprinted 1996 (twice with corrections)
Reprinted 1997 (twice), 1998

Printed in the United States of America

Typeset in TeX

A catalogue record for this book is available from the British Library

Library of Congress Cataloguing-in-Publication Data is available

ISBN 0-521-49699-3 hardback
ISBN 0-521-49789-2 paperback

Contents

Preface

'Finance' is one of the fastest developing areas in the modern banking and corporate world. This, together with the sophistication of modern financial products, provides a rapidly growing impetus for new mathematical models and modern mathematical methods; the area is an expanding source for novel and relevant 'real-world' mathematics. The demand from financial institutions for well-qualified mathematicians is substantial, and there is a corresponding need for professional training of existing staff. Since 1992 the authors of this book have, in response, given graduate and undergraduate level courses on the subject. We have also organised a series of professional development courses for practitioners, held in Oxford and New York, with the assistance of Oxford University's Department for Continuing Education and the Oxford Centre for Industrial and Applied Mathematics. The material and notes from these courses became a book, *Option Pricing: Mathematical Models and Computation*, an advanced yet accessible account of applied and numerical techniques in the area of derivatives pricing.

Following the success of *Option Pricing* among financial practitioners, we have written this student-oriented version as an introduction to the subject. Our aim in *The Mathematics of Financial Derivatives: A Student Introduction* is to introduce the principles in a clear and readable way while leaving the more advanced topics and detailed practicalities, especially numerical issues, to the earlier book.

In what follows we describe the modelling of financial derivative products from an applied mathematician's viewpoint, from modelling through analysis to elementary computation. Some mathematics is assumed, but we explain everything that is not contained in the early calculus, probability and algebra courses of an undergraduate degree or equivalent in mathematics, physics, chemistry, engineering or similar subjects. We also give enough detail of the finance that the book can be read by mathematicians whose knowledge of financial markets is only sketchy. It is sufficiently self-contained that it could be used for a course on the subject, on its own or in conjunction either with a more probability-based text such as Duffie (1992) or with a more practically oriented book such as Hull (1993) or Gemmill (1992).

Our philosophy may be described briefly as follows:

- We present a unified approach to modelling many derivative products as partial differential equations. We make no more fuss over valuing an average strike option (a particularly exotic product) than over valuing the simplest option. There is a minimal use of fudges or approximations.
- We describe the theory of partial differential equations. We explain why they are one of the best approaches to modelling in many physical or financial subjects.
- We are happy to use numerical solutions. We would rather have an accurate numerical solution of the correct model than an explicit solution of the wrong model.

The authors of any book on financial models must decide at the outset what will be the mathematical basis of their approach. Essentially, this entails a decision on the amount of more or less rigorous analysis to incorporate, along with a choice whether or not to couch the discussion largely in terms of the language of stochastic processes. We feel that the interests of communication with our readers, especially those at the practical end of the subject, are best served by a relatively informal approach. We have therefore tried to stress the intuitive aspects of the subject, and this has led us naturally to emphasise the derivation and use of differential equations and associated numerical techniques. We hope that the rigour thereby forgone is compensated for by improved directness. We emphasise that there are excellent texts that fully cover more theoretical aspects of the subject.

The first chapter is an introduction to the subject of option pricing and markets. It is aimed at the reader who is new to the financial side of the subject and it contains no technical mathematical material.

Chapter 2 opens with a description of the movement of asset prices as a random walk. The underlying basis for the models used thereafter is that the future value of asset prices cannot be predicted with certainty. This does not mean that the movements cannot be modelled, only that any such model must incorporate a degree of randomness. We see that, despite the random nature of asset prices, there are many problems for which we can make deterministic (that is, not probabilistic) statements. It is fortunate that such problems also happen to be the most interesting and important financially. Later in this chapter we informally describe the methods of stochastic calculus, building intuitively on the ideas of ordinary calculus.

In Chapter 3 the mathematical modelling becomes more explicitly related to option pricing. This chapter is perhaps the most important in the book. In it we present the cornerstone of the subject of option pricing: the derivation of the original Black–Scholes partial differential equation and boundary conditions for the value of an option.

The fact that partial differential equations prove to be central to our approach to financial modelling provides the motivation for the next four chapters. The type of partial differential equation that occurs most often in financial theory is the parabolic partial differential equation, the canonical example of which is the heat or diffusion equation. Posing the problems in the form of a parabolic partial differential equation means that we have nearly two centuries' worth of theory on which to call. Once the problem has been presented in such a form we may consider ourselves to be on well-known territory.

In Chapter 4 we discuss linear parabolic second order partial differential equations in quite general terms. In Chapter 5 the diffusion equation is dealt with in some detail. The general solution of the initial value problem on the whole line is derived and used to deduce the Black–Scholes formulæ for European call and put options. In Chapter 6 we discuss modifications necessary to account for the payment of dividends and then derive explicit formulæ for option prices when parameters are time-dependent. We also analyse futures and forward contracts, and options on them. The first part of the book concludes in Chapter 7 with a discussion of the free boundary problems that arise from models of American options where the possibility of early exercise gives rise to a free boundary. The theory of this chapter sets the scene for the numerical solution of free boundary problems.

Many of the models we derive do not admit closed form solutions. While it might be possible to modify them so that they can be solved in closed form, such modifications would probably have no basis in financial reality. We prefer, therefore, to accept that, for many practical problems, numerical methods of solution are necessary. The second part of the book is devoted to this topic.

In Chapter 8, we introduce finite differences for continuous models of

European options. We demonstrate the explicit finite-difference method in some detail, and similarly describe the fully implicit and Crank–Nicolson methods. In Chapter 9 we discuss the numerical solution of free boundary problems for American options, again with particular emphasis on finite-difference methods. The final chapter on numerical methods, Chapter 10, gives details of discrete binomial models, which are a popular, albeit limited, alternative to finite differences.

Having dealt with the necessary basic theory, we come to some more advanced subjects in mathematical finance in the remainder of the book. Part 3 deals with so-called exotic and path-dependent options, and with the influence of transaction costs. We begin in Chapter 11 with an overview of exotic options, and we describe in more detail some quite simple contracts. Chapter 12 deals with another straightforward extension, this time to barrier options. In Chapter 13 we derive a general theory for options depending on history functionals of asset prices and give several examples in detail. Two path-dependent options have chapters to themselves: Asian options, which involve an average of the asset price, in Chapter 14, and lookback options, depending on the realised maximum or minimum, in Chapter 15. Finally, in Chapter 16 we consider a simple model for options in the presence of transaction costs. These can affect option prices quite significantly, and the model we discuss is of both practical importance and mathematical interest.

Throughout the first three parts of the book, the only random variable in our problems is the asset price. In the last part, we allow the interest rate to be unpredictable. Chapter 17 deals with the pricing of bonds and other interest rate derivatives; this entails the introduction of a simple stochastic model for the short-term interest rate. We conclude the book in Chapter 18 with a brief discussion of the valuation of convertible bonds; these bonds can have a value that depends on *two* random variables, an underlying asset and an interest rate.

Many people have offered us help, advice and encouragement during the production of both this book and *Option Pricing*. We are grateful in particular to the delegates on our courses, who gave us valuable insight into the workings of financial markets and made many helpful suggestions about the subjects of practical interest; to the graduate students and colleagues at the Oxford Centre for Industrial and Applied Mathematics; to the staff at the Oxford Centre for Continuing Professional Development; and lastly to the staff of Oxford Financial Press and Cambridge University Press for their continuing encouragement.

The success of mathematics in finance depends heavily on the contributions of researchers in universities and financial institutions. It is intended that part of the royalties from the sales of this book will be used to fund a graduate scholarship for outstanding students who wish to study for a doctorate in the subject at the Oxford Centre for Industrial Applied Mathematics, Oxford University. For details of this award, which is intended as a supplement to the usual costs of fees and maintenance, write to: The Administrator, OCIAM, Mathematical Institute, 24–29 St. Giles', Oxford, OX1 3LB, UK.

Technical Point

Throughout the book, at the end of many sections and subsections, are scattered 'Technical Points'. As the name suggests, these items describe some of the more technical matters in our subject, matters which would disrupt the flow if contained in the main body of the text, yet which are too large to appear as footnotes. These items may be ignored on a first reading.

Occasionally, some words or phrases appear in **bold face**. This means that such a word or phrase is being defined. In some cases this definition is technical and in others simply descriptive.

Further Reading

There is a huge literature on financial mathematics. We have given selected references at the end of each chapter. Much of the material in the present book, especially in the chapters on numerical methods and exotic options, is covered in greater detail in *Option Pricing: Mathematical Methods and Computation*, also written by the present authors. This book is available from the publishers, Oxford Financial Press, PO Box 348, Oxford OX4 1DR, UK.

Exercises

Exercises are provided at the end of each chapter. Some of them are quite closely based on the text, while others, particularly in the later chapters, should be regarded as invitations to experiment. Hints to selected exercises are given at the end of the book.

Part one

Basic Option Theory

1 An Introduction to Options and Markets

1.1 Introduction

This book is about mathematical models for financial markets, the assets that are traded in them and, especially, financial derivative products such as options and futures. There are many kinds of financial market, but the most important ones for us are:

- **Stock markets**, such as those in New York, London and Tokyo;
- **Bond markets**, which deal in government and other bonds;
- **Currency markets** or **foreign exchange markets**, where currencies are bought and sold;
- **Commodity markets**, where physical assets such as oil, gold, copper, wheat or electricity are traded;
- **Futures** and **options** markets, on which the derivative products that are the subject of this book are traded.

The reader may not have encountered all of the financial terms in bold face in this list. Most will be explained in detail later in the book when we need them. However, we do assume that the *raison d'être* of the currency and commodity markets is clear, and we hope that readers are familiar with the idea behind **stocks** (also known as **shares** or **equities**). Roughly speaking, a company that needs to raise money, for example to build a new factory or develop a new product, can do so by selling shares in itself to investors. The company is then 'owned' by its shareholders; if the company makes a profit, part of this may be paid out to shareholders as a **dividend** of so much per share, and if the company is taken over or otherwise wound up, the proceeds (if any) are distributed to shareholders. Shares thus have a value that reflects the views of investors about the likely future dividend payments and capital growth of the company; this value is quantified by the price at which they are bought and sold on stock exchanges.[1]

We have, then, a collection of markets on which assets of various kinds are bought and sold. As markets have become more sophisticated, more complex contracts than simple buy/sell trades have been introduced. Known as **financial derivatives**, **derivative securities**, **derivative products**, **contingent claims** or just **derivatives**, they can give investors of all kinds a great range of opportunities to tailor their dealings to their investment needs. This book explains some of the financial theory and models that have been developed to analyse derivatives, a theory that is necessarily mathematical in character (the specialists now employed by all major financial institutions to work in this area are called 'rocket scientists'!), but which is at bottom a very elegant and clear combination of mathematical modelling and analysis. First, though, we need to become familiar with some of the necessary financial jargon, and to see how derivatives work. We begin with the example of an option, which is one of the commonest examples of a derivative security.

1.2 What is an Option?

The simplest financial option, a **European call option**, is a contract with the following conditions:

- At a prescribed time in the future, known as the **expiry date** or **expiration date**, the **holder** of the option *may*
- purchase a prescribed asset, known as the **underlying asset** or, briefly, the **underlying**, for a
- prescribed amount, known as the **exercise price** or **strike price**.

The word 'may' in this description implies that for the holder of the option, this contract is a *right* and not an *obligation*. The other party to the contract, who is known as the **writer**, does have a potential obligation: he *must* sell the asset if the holder chooses to buy it. Since the option confers on its holder a right with no obligation it has some value. Moreover, it must be paid for at the time of opening the contract. Conversely, the writer of the option must be compensated for the obligation he has assumed. Two of our main concerns throughout this book are:

[1] In practice, companies may have a much more complex structure for their equity, but in an introductory text we try not to get enmeshed in these details. The ideas we describe carry over with the appropriate modifications throughout.

- How much would one pay for this right, i.e. what is the value of an option?
- How can the writer minimise the risk associated with his obligation?

A Simple Example: A Call Option

How much is the following option now worth? Today's date is 22 August 1995.

- On 14 April 1996 the holder of the option *may*
- purchase one XYZ share for 250p.

In order to gain an intuitive feel for the price of this option let us imagine two possible situations that might occur on the expiry date, 14 April 1996, nearly eight months in the future.

If the XYZ share price is 270p on 14 April 1996, then the holder of the option would be able to purchase the asset for only 250p. This action, which is called **exercising** the option, yields an immediate profit of 20p. That is, he can buy the share for 250p and immediately sell it in the market for 270p:

$$270\text{p} - 250\text{p} = 20\text{p} \ \ \text{profit}.$$

On the other hand, if the XYZ share price is only 230p on 14 April 1996 then it would not be sensible to exercise the option. Why buy something for 250p when it can be bought for 230p elsewhere?

If the XYZ share only takes the values 230p or 270p on 14 April 1996, with equal probability, then the expected profit to be made is

$$\tfrac{1}{2} \times 0 \ + \ \tfrac{1}{2} \times 20 = 10\text{p}.$$

Ignoring interest rates for the moment, it seems reasonable that the order of magnitude for the value of the option is 10p.

Of course, valuing an option is not as simple as this, but let us suppose that the holder did indeed pay 10p for this option. Now if the share price rises to 270p at expiry he has made a net profit calculated as follows:

$$\begin{array}{rcr} \text{profit on exercise} & = & 20\text{p} \\ \text{cost of option} & = & -10\text{p} \\ \hline \text{net profit} & = & 10\text{p} \\ \hline \end{array}$$

This net profit of 10p is 100% of the up-front premium. The downside of this speculation is that if the share price is less than 250p at expiry he has lost all of the 10p invested in the option, giving a loss of 100%. If the investor had instead purchased the share for 250p on 22 August 1995, the corresponding profit or loss of 20p would have been only ±8% of the original investment. Option prices thus respond in an exaggerated way to changes in the underlying asset price. This effect is called **gearing**.

We can see from this simple example that the greater the share price on 14 April 1996, the greater the profit. Unfortunately, we do not know this share price in advance. However, it seems reasonable that the higher the share price is now (and this is something we *do* know) then the higher the price is likely to be in the future. Thus the value of a call option *today* depends on today's share price. Similarly, the dependence of the call option value on the exercise price is obvious: the lower the exercise price, the less that has to be paid on exercise, and so the higher the option value.

Implicit in this is that the option is to expire a significant time in the future. Just before the option is about to expire, there is little time for the asset price to change. In that case the price at expiry is known with a fair degree of certainty. We can conclude that the call option price must also be a function of the time to expiry.

Later we also see how the option price depends on a property of the 'randomness' of the asset price, the volatility. The larger the volatility, the more jagged is the graph of asset price against time. This clearly affects the distribution of asset prices at expiry, and hence the expected return from the option. The value of a call option should therefore depend on the volatility. Finally, the option price must depend on prevailing bank interest rates; the option is usually paid for up-front at the opening of the contract whereas the payoff, if any, does not come until later. The option price should reflect the income that would otherwise have been earned by investing the premium in the bank.

Put Options

The option to *buy* an asset discussed above is known as a **call** option. The right to *sell* an asset is known as a **put** option and has payoff properties which are opposite to those of a call. A put option allows its holder to sell the asset on a certain date for a prescribed amount. The writer is then obliged to buy the asset. Whereas the holder of a call option wants the asset price to rise – the higher the asset price at expiry the greater the profit – the holder of a put option wants the asset price to fall as low as possible. The value of a put option also increases with the exercise price, since with a higher exercise price more is received for the asset at expiry.

1.3 Reading the Financial Press

Armed with the jargon of calls, puts, expiry dates and so forth, we are in a position to read the options pages in the financial press. Our examples are taken from the *Financial Times* of Thursday 4 February 1993.

In Figure 1.1 is shown the traded[1] options section of the *Financial Times.* This table shows the prices of some of the options traded on the London International Financial Futures and Options Exchange (LIFFE). The table lists the last quoted prices on the previous day for a large number of options, both calls and puts, with a variety of exercise prices and expiry dates. Most of these examples are options on individual equities, but at the bottom of the third column we see options on the *FT-SE* index, which is a weighted arithmetic average of 100 equity shares quoted on the London Stock Exchange.

First, let us concentrate on the prices quoted for Rolls-Royce options, to be found in the third column labelled 'R. Royce'. Immediately beneath R. Royce is the number 134 in parentheses. This is the closing price, in pence, of Rolls-Royce shares on the previous day. To the right of R. Royce/(134) are the two numbers 130 and 140: these are two exercise prices, again in pence. Note that for equity options the *Financial Times* prints only those exercise prices each side of the closing price. Many other exercise prices exist (at intervals of 10p in this case) but are not printed in the *Financial Times* for want of space.

Now examine the six numbers to the right of the 130. The first three (11, 15, 19) are the prices of call options with different expiry dates, and the next three (9, 14, 17) are the prices of put options. The expiry date of each of these options can be found by looking at the top of its column. There we see that there are options on Rolls-Royce shares expiring in March, June and September, at a specified time on a specified date in each month, in this case at 18:00 on the third Wednesday of the month concerned (trading ceases slightly earlier). Option prices are quoted on an exchange only for a small number of expiry dates and only for exercise prices at discrete intervals (here ..., 130, 140, ...). For LIFFE-traded options on equities the expiry dates come in intervals of three months. When it is created, the longest dated option has a lifespan

[1] The word 'traded' here refers to an option that is traded on an exchange such as LIFFE or the CBOE (Chicago Board Options Exchange).

LIFFE EQUITY OPTIONS

Option		CALLS Apr	Jul	Oct	PUTS Apr	Jul	Oct
Alld Lyons	550	48	60	68	10	24	30
(°585)	600	22	34	44	34	50	55
ASDA	57	9	12½	14	4	7	9
(°61)	67	4	8	9	8½	11½	14
Brit Airways	280	27	33	39	9½	20	24
(°296)	300	17	24	29	18	30	34
SmKl Bchm A	460	30	47	53	17	26	52
(°470)	500	13	27	36	42	48	54
Boots	500	23	32	43	21	33	37
(°501)	550	6½	15	23	56	67	70
B P	260	16½	23	29	11	17	21
(°265)	280	8	15	19	22	28	34
British Steel	70	13	16½	19	3½	6½	8½
(°79)	80	8	12	14	7½	11½	13
Bass	600	31	48	61	24	38	45
(°604)	650	12	27	39	58	70	73
C & Wire	700	40	60	70	25	43	50
(°710)	750	18	35	46	56	73	80
Courtaulds	550	40	52	61	17	30	37
(°568)	600	16	29	38	45	60	64
Com. Union	600	35	49	57	21	32	40
(°623)	650	11	26	34	53	61	70
Fisons	220	22	31	39	18	30	35
(°222)	240	14	22	31	31	43	48
GKN	460	27	36	45	21	28	34
(°472)	500	8½	20	28	48	53	57
Grand Met	420	38	53	59	11	21	27
(°439)	460	16	31	36	30	41	47
I.C.I	1100	53	82	92	50	68	88
(°1132)	1150	31	62	72	80	100	117
Kingfisher	550	34	48	53	21	38	45
(°559)	600	12	25	33	52	68	73
Ladbroke	200	16	25	29	18	26	32
(°202)	220	8	16	22	32	38	44
Land Secur	460	45	49	53	5	17	21
(°493)	500	17	25	31	20	38	41
M & S	330	18	25	34	12	20	24
(°333)	360	6	12	20	32	38	41
Sainsbury	550	43	54	63	12	24	29
(°577)	600	16	28	38	38	50	56
Shell Trans	550	32	44	50	12	19	26
(°576)	600	6	19	25	44	47	53
Storehouse	200	18	26	34	8	17	18
(°204)	220	11	17	22	21	25	28
Trafalgar	90	11	14	18	8	9	13
(°93)	100	6½	11	14	12	17	18
Utd Biscuits	360	18	25	33	19	25	29
(°366)	390	6	14	20	41	45	49
Unilever	1100	72	90	110	16	35	42
(°1149)	1150	39	60	82	42	56	63

Option		Feb	May	Aug	Feb	May	Aug
Brit Aero	280	23	40	53	16	34	46
(°287)	300	14	33	47	27	49	60

Option		CALLS Feb	May	Aug	PUTS Feb	May	Aug
BAA	750	41	61	71	6	16	31
(°786)	800	11	33	45	29	41	55
BAT Inds	950	45	58	74	8	37	46
(°982)	1000	16	33	50	30	64	73
BTR	550	22	30	38	5½	20	28
(°565)	600	3	10	19	38	53	55
Brit Telecom	420	10	23	29	7	15	24
(°420)	460	1	8	12½	40	41	50
Cadbury Sch	460	15	25	35	8	24	30
(°465)	500	3½	10	19	37	51	55
Eastern Elec	400	18	32	-	6	15	-
(°411)	430	5	-	-	24	-	-
Guinness	460	24	36	46	8	25	30
(°473)	500	6	19	28	33	48	56
GEC	300	9	19	23	7	13	21
(°301)	330	1	7	11	30	32	41
Hanson	260	7½	14½	18	5½	12	16½
(°262)	280	1	5½	9½	20	25	28
LASMO	160	8	18	22	9	18	23
(°161)	180	3	8½	15	24	33	36
Lucas Inds	140	15	23	26	4	11	16
(°151)	160	4	13	17	14	24	28
P. & O	550	24	42	52	15	40	52
(°564)	600	5½	21	33	45	72	82
Pilkington	100	7	15	18	6	12	16
(°102)	110	4	10	14	12	17	22
Prudential	300	24	30	34	2½	11	15
(°321)	330	6	13	19	14	26	31
R.T.Z	650	34	47	60	9	30	40
(°672)	700	10	25	39	36	59	69
Scot. & New	420	23	38	45	5	13	24
(°343)	460	4½	17	24	26	35	48
Tesco	240	22	27	30	2	8	12
(°257)	260	7	14	21	9	19	22
Thames Wtr	460	24	37	42	3½	12	22
(°479)	500	3½	16	20	25	32	45
Vodafone	390	18	34	43	8	21	28
(°398)	420	5½	19	29	27	36	45

Option		Mar	Jun	Sep	Mar	Jun	Sep
Abbey Nat	360	28	34	42	11	17	23
(°379)	390	11	19	27	27	33	38
Amstrad	20	5½	6½	7½	1½	1½	2½
(°24)	25	2½	3½	4½	3½	4	4½
Barclays	420	45	54	59	11	21	30
(°458)	460	20	33	38	33	41	51
Blue Circle	220	22	30	35	10	21	27
(°230)	240	12	19	27	22	33	39
British Gas	280	15½	19	23	6½	15	18
(°287)	300	6½	9½	14	18	27	30
Dixons	220	15	25	28	13	19	25
(°221)	240	8	17	21	26	31	38
Eurotunnel	420	38	55	70	20	35	45
(°435)	460	18	37	50	45	57	67

Option		CALLS Mar	Jun	Sep	PUTS Mar	Jun	Sep
Glaxo	650	38	60	82	29	42	52
(°664)	700	17	38	58	60	72	80
Hillsdown	140	17	22	24	6	15	20
(°147)	160	6	12	16	18	27	31
Lonrho	70	9	12	15	5	8½	11
(°75)	80	4½	8	11½	11	14	17
HSBC 75p shs	550	45	57	70	17	34	46
(°576)	600	18	32	48	44	60	70
Natl Power	280	24	32	36	6	11	18
(°294)	300	11	21	25	15	20	27
Reuters	1400	69	108	135	47	78	100
(°1426)	1450	40	82	113	78	103	122
R. Royce	130	11	15	19	9	14	17
(°134)	140	6	11	16	16	20	23
Scot Power	200	20	25	28	2	5	10½
(°216)	220	6	13	17	9	13½	22
Sears	100	9	11	15	5	8½	11
(°104)	110	4	7	9½	10	14	16
Forte	180	15	20	25	10	19	24
(°186)	200	7	13	17	21	32	36
Thorn EMI	800	61	81	90	10	22	39
(°843)	850	28	50	60	29	43	66
TSB	160	19	23	27	5	8½	12
(°176)	180	6½	13	17	15	18	22
Vaal Reefs	30	6	6	6½	3	4	4½
(°$34)	35	2	4	5½	5½	6	8
Wellcome	850	50	75	100	28	48	63
(°866)	900	26	52	75	57	77	92

EURO FT-SE INDEX (°2872)

	2675	2725	2775	2825	2875	2925	2975	3025
CALLS								
Feb	205	155	108	68	36	16	5	2
Mar	215	170	129	92	61	38	22	11
Apr	-	188	-	115	-	61	-	28
Jun	-	217	-	152	-	100	-	60
Sep	-	252	-	188	-	129	-	-
PUTS								
Feb	1½	3½	7	15	34	65	104	151
Mar	9	12	21	34	53	81	114	153
Apr	-	26	-	51	-	97	-	163
Jun	-	50	-	80	-	125	-	185
Sep	-	-	-	120	-	-	-	-

FT-SE INDEX (°2872)

	2650	2700	2750	2800	2850	2900	2950	3000
CALLS								
Feb	233	183	135	89	50	24	9	3
Mar	243	197	153	113	79	51	31	17
Apr	254	210	170	131	99	73	51	36
May	266	226	186	151	121	93	72	52
Jun	-	235	-	164	-	107	-	67
Dec †	-	-	-	235	-	187	-	130
PUTS								
Feb	1½	2½	4½	10	23	47	85	131
Mar	8	12	17	29	46	69	100	138
Apr	16	22	31	44	63	86	116	151
May	26	35	45	61	80	105	133	166
Jun	-	40	-	70	-	113	-	175
Dec †	-	90	-	125	-	185	-	-

February 3 Total Contracts 31,257
Calls 21,861 Puts 9,396
FT-SE Index Calls 7,946 Puts 4,410
Euro FT-SE Calls 816 Puts 278
°Underlying security price † Long dated expiry mths
Premiums shown are based on closing offer prices

Figure 1.1 The traded options section of the *Financial Times* of 4 February 1993.

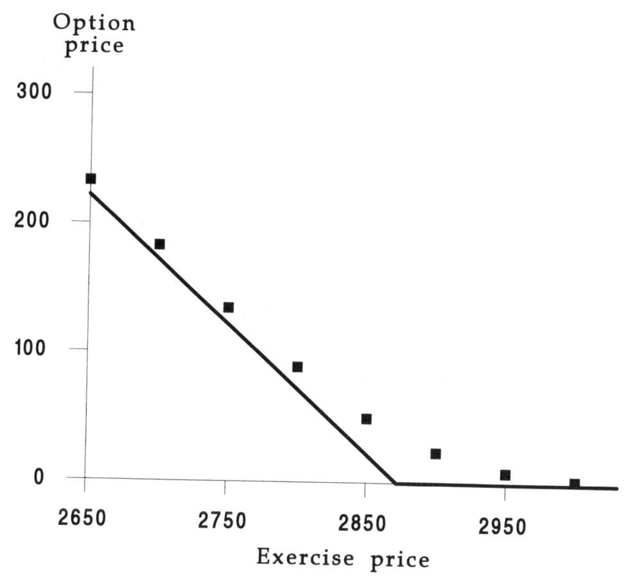

Figure 1.2 The *FT-SE* index call option values versus exercise price and the option values at expiry assuming that the index value is then 2872.

of nine months. Later in the year the December series of Rolls-Royce options will come into being.

Since a call option permits the holder to pay the exercise price to obtain the asset, we can see that call options with exercise price 140p are cheaper than those with exercise price 130p. This is because more must be paid for the share at exercise. The converse is true for puts: the holder of a 140p put can realise more by selling the share at exercise than the holder of a 130p put, and so the former is worth more.

Now let us look at the options on the *FT-SE* index. Towards the bottom of the third column we see prices for the *FT-SE* index call options. (Although the index is just a number, the contract is given a nominal price in pounds equal to 10 times the *FT-SE* value.) The exercise prices are quoted at intervals of 50 from 2650 to 3000 and expiry dates at monthly intervals. Since these options expire on the third Friday of the month, the February options have only about 10 days left. In Figure 1.2 we plot the value of the February call options against exercise price.

The closing value of the *FT-SE* index on 3 February 1993 was 2872. Suppose that the *FT-SE* index had exactly the same value at expiry as on 3 February 1993. Then the value of each call option contract at expiry would be the 'ramp function'

$$\begin{array}{ll} £10 \times (2872 - \text{exercise value}) & \text{for exercise value} \leq 2872 \\ 0 & \text{for exercise value} \geq 2872. \end{array}$$

In Figure 1.2 we also plot this ramp function. Notice that the data points are close to but above the ramp function. The difference between the two is due to the indeterminacy in the future index value: the index is unlikely to be at 2872 at the time of expiry of the February options. We return to the example of the *FT-SE* index call options in Chapter 3.

Finally, note that for each option type there is only one quoted price in this table. In reality the option could not be bought and sold for the same price since the market-maker has to make a living. Thus there are *two* prices for the option. The investor pays the ask (or offer) price and sells for the bid price, which is less than the ask price. The price quoted in the newspapers is usually a mid-price, the average of the bid and ask prices. The difference between the two prices is known as the bid-ask or bid-offer spread.

Technical Point: The trading of options.

Before 1973 all option contracts were what is now called 'over-the-counter' (OTC). That is, they were individually negotiated by a broker on behalf of two clients, one being the buyer and the other the seller. Trading on an official exchange began in 1973 on the Chicago Board Options Exchange (CBOE), with trading initially only in call options on some of the most heavily traded stocks. As increased competition followed the listing of options on an exchange, the cost of setting up an option contract decreased significantly.

Options are now traded on all of the world's major exchanges. They are no longer restricted to equity options but include options on indices, futures, government bonds, commodities, currencies etc. The OTC market still exists, and options are written by institutions to meet a client's needs. This is where exotic option contracts are created; they are very rarely quoted on an exchange.

When an option contract is initiated there must be two sides to the agreement. Consider a call option. On one side of the contract is the buyer, the party who has the right to exercise the option. On the other side is the party who must, if required, deliver the underlying asset. The latter is called the **writer** of the option.

Many options are registered and settled via a **clearing house**. This central body is also responsible for the collection of **margin** from the writers of options. This margin is a sum of money (or equivalent) which is held by the clearing house on behalf of the writer. It is a guarantee that he is able to meet his obligations should the asset price move against him.

The trade in the simplest call and put options (colloquially called **vanilla** options, because they are ubiquitous) is now so great that it can, in some markets, have a value in excess of that of the trade in the underlying. In some cases too the exchange-traded options are more liquid than the underlying asset. To give an idea of the size[3] of the derivatives (including futures) markets, there is an estimated $10,000 *billion* in derivatives investments worldwide in total (this is a gross figure; the net figure is much smaller). In late 1992, Citicorp alone had an estimated exposure equivalent to a notional contract value of $1426bn. As the number and type of derivative products have increased so there has been a corresponding growth in option pricing as a subject for academic and corporate research. This is especially true today as increasingly exotic types of options are created.

1.4 What are Options For?

Options have two primary uses: speculation and hedging. An investor who believes that a particular stock, XYZ again, say, is going to rise can purchase some shares in that company. If he is correct, he makes money, if he is wrong he loses money. This investor is speculating. As we have noted, if the share price rises from 250p to 270p he makes a profit of 20p or 8%. If it falls to 230p he makes a loss of 20p or 8%. Alternatively, suppose that he thinks that the share price is going to rise within the next couple of months and that he buys a call with exercise price 250p and expiry date in three months' time. We have seen in the earlier example that if such an option costs 10p then the profit or loss is magnified to 100%. Options can be a cheap way of exposing a portfolio to a large amount of risk.

If, on the other hand, the investor thinks that XYZ shares are going to fall he can, conversely, sell shares or buy puts. If he speculates by selling shares that he does not own (which in certain circumstances is perfectly legal in many markets) he is selling **short** and will profit from a fall in XYZ shares. (The opposite of a short position is a **long** position.) The

[3] These values are taken from a review of the derivatives market in the *Financial Times* of 8 December 1992.

same argument concerning the exaggerated movement of option prices applies to puts as well as calls, and if he wants to speculate he may decide to buy puts instead of selling the asset. However, suppose that the investor already owns XYZ shares as a long-term investment. In this case he might wish to insure against a temporary fall in the share price, while being reluctant to liquidate his XYZ holdings only to buy them back again later, possibly at a higher price if his view of the share price is wrong, and certainly having incurred some transaction costs on the two deals.

The discussion so far has been from the point of view of the holder of an option. Let us now consider the position of the other party to the contract, the writer. While the holder of a call option has the possibility of an arbitrarily large payoff, with the loss limited to the initial premium, the writer has the possibility of an arbitrarily large *loss*, with the profit limited to the initial premium. Similarly, but to a lesser extent, writing a put option exposes the writer to large potential losses for a profit limited to the initial premium. One could therefore ask

- Why would anyone write an option?

The first likely answer is that the writer of an option expects to make a profit by taking a view on the market. Writers of calls are, in effect, taking a short position in the underlying: they expect its value to fall. It is usually argued that such people must be present in the market, for if everyone expected the value of a particular asset to rise its market price would be higher than, in fact, it is. (These 'bears' are also potential customers for put options on the underlying.) Similarly, there must also be people who believe that the value of the underlying will rise (or the price would be lower than, in fact, it is). These 'bulls' are potential writers of put options and buyers of call options. An extension of this argument is that writers of options are using them as insurance against adverse movements in the underlying, in the same way as holders do.

Although this motivation is plausible, it is not the whole story. Market makers have to make a living, and in doing so they cannot necessarily afford to bear the risk of taking exposed positions. Instead, their profit comes from selling at slightly above the 'true value' and buying at slightly below; the less risk associated with this policy, the better. This idea of reducing risk brings us to the subject of **hedging**. We introduce it by a simple example.

Since the value of a put option rises when an asset price falls, what happens to the value of a portfolio containing both assets and puts?

The answer depends on the ratio of assets and options in the portfolio. A portfolio that contains only assets falls when the asset price falls, while one that is all put options rises. Somewhere in between these two extremes is a ratio at which a small unpredictable movement in the asset does not result in any unpredictable movement in the value of the portfolio. This ratio is instantaneously risk-free. The reduction of risk by taking advantage of such correlations between the asset and option price movements is called **hedging**. *If a market maker can sell an option for more than it is worth and then hedge away all the risk for the rest of the option's life, he has locked in a guaranteed, risk-free profit.* This idea is central to the theory and practice of option pricing.

Beyond the primary roles just discussed, many more general problems can be cast in terms of options. This is an increasingly important way of analysing decision-making. A simple example is that of a company which owns a mine, from which gold can be produced at a known cost. The mine can be started up and closed down, depending on current gold prices. How much does this flexibility add to the value of the company in the eyes of a predator, or of its shareholders? An answer can be arrived at by modelling the mine as an option, in this case on gold. In a similar vein, in valuing of a piece of vacant land we may want to try to quantify the value added by the fact that property prices may rise fast enough in the future for it to be worth leaving the land vacant for later resale.

1.5 Other Types of Option

Call and put options form a small section of the available derivative products. Our earlier description of an option contract concentrated on a European option, but nowadays most options are what is called American. The European/American classification has nothing to do with the continent of origin but refers to a technicality in the option contract. An **American option** is one that may be exercised at *any time* prior to expiry. The options described above, which may only be exercised *at* expiry, are called **European**. To the mathematician, American options are more interesting since they can be interpreted as free boundary problems – we see this in Chapter 7 and again in Chapter 9. Not only must a *value* be assigned to the option but, and this is a feature of American options only, we must determine *when it is best to exercise the option*. We see that the 'best' time to exercise is not subjective, but that it can be determined in a natural and systematic way.

Other types of option which we describe in this book include the so-called exotic or path-dependent options. These options have values which depend on the history of an asset price, not just on its value on exercise. An example is an option to purchase an asset for the arithmetic average value of that asset over the month before expiry. An investor might want such an option in order to hedge sales of a commodity, say, which occur continually throughout this month. Another example might be an oil refiner who buys oil at the spot rate, which may vary, but wants to sell the refined product at a constant price. Once the idea of history dependence is accepted it is a very small step to imagining options which depend on the geometric average of the asset price, the maximum or the minimum of the asset price, etc. This then brings us to the question of how to calculate the arithmetic average, say, of an asset price which may be quoted every 30 seconds or so; for a very liquid stock this would give about 250,000 prices per year. In practice the option contract might specify that the arithmetic average is the mean of the closing price every business day, of which there are only 250 every year. (In contrast to 'tick data', these latter prices are reliable and not open to dispute.) Does this 'discrete sampling' give different option values if the sampling takes place at different times?

We show how to put the following options into a unifying framework:

- barrier options (the option can either come into existence or become worthless if the underlying asset reaches some prescribed value before expiry);
- Asian options (the price depends on some form of average);
- lookback options (the price depends on the asset price maximum or minimum).

We discuss European and American versions of these as well as both continuous and discrete sampling of the history-dependent factor.

1.6 Forward and Futures Contracts

Apart from options, we shall also analyse two other common contingent claims, forward contracts and futures contracts. A **forward contract** is an agreement between two parties whereby one contracts to buy a specified asset from the other for a specified price, known as the **forward price**, on a specified date in the future, the **delivery date** or **maturity**

date. This contract has similarities to an option contract if we think of the forward price as equivalent to the exercise price. However, what is lacking is the element of choice: the asset *has* to be delivered and paid for. A forward contract is also different from an option contract in that no money changes hands until delivery, whereas the premium for an option is paid up-front. It therefore costs nothing to enter into a forward contract. A further difference from option contracts is that the forward price is not set at one of a number of fixed values for all contracts on the same asset with the same expiry. Instead, it is determined at the outset, individually for each contract.

A **futures contract** is in essence a forward contract, but with some technical modifications. Whereas a forward contract may be set up between any two parties, futures are usually traded on an exchange which specifies certain standard features of the contract such as delivery date and contract size. A further complication is the margin requirement, a system designed to protect both parties to a futures contract against default. Whereas the profit or loss from a forward contract is only realised at the expiry date, the value of a futures contract is evaluated every day, and the change in value is paid to one party by the other, so that the net profit or loss is paid across gradually over the lifetime of the contract. Despite these differences, it can be shown that under some not too restrictive assumptions the futures price is almost the same as the forward price. When interest rates are predictable, the two coincide exactly. For later use, we note that it again costs nothing to enter into a futures contract.

Because neither forward nor futures contracts contain the element of choice (to exercise or not to exercise) that is inherent in an option, they are easier to value. Nevertheless, because they are not central to our development of the subject, we defer their treatment until Chapter 6.

1.7 Interest Rates and Present Value

For almost the whole of this book we assume that the short-term bank deposit interest rate is a known function of time, not necessarily constant. This is not an unreasonable assumption when valuing options, since a typical equity option has a total lifespan of about nine months. During such a relatively short time interest rates may change but not usually by enough to affect the prices of options significantly. (An interest rate change from 8% *p.a.* to 10% *p.a.* typically decreases a nine-month option value by about 2%.) However, towards the end of the book, in Chapters 17 and 18 on bond pricing, we relax the assumption of known interest rates and present a model where the short-term rate is a random variable. This is important in valuing interest rate dependent products, such as bonds, since they have a much longer lifespan, typically 10 or 20 years; the assumption of known or constant interest rates is not a good one over such a long period.

For valuing options the most important concept concerning interest rates is that of **present value** or **discounting**. Ask the question

- How much would I pay *now* to receive a guaranteed amount E at the future time T?

If we assume that interest rates are constant, the answer to this question is found by discounting the future value, E, using continuously compounded interest. With a constant interest rate r, money in the bank $M(t)$ grows exponentially according to

$$\frac{dM}{M} = r\,dt. \tag{1.1}$$

The solution of this is simply

$$M = ce^{rt},$$

where c is the constant of integration. Since $M = E$ at $t = T$, the value at time t of the certain payoff is

$$M = Ee^{-r(T-t)}.$$

If the interest rate is a known function of time $r(t)$, then (1.1) can be modified trivially and results in

$$M = Ee^{-\int_t^T r(s)ds}.$$

Further Reading

- Sharpe (1985) describes the workings of financial markets in general. It is a very good broad introduction to investment theory and practice.
- Blank, Carter & Schmiesing (1991) discuss the uses of options and other products by different sorts of finance practitioners. Copeland, Koller & Murrin (1990) discuss the use of options in valuing companies.

- Good descriptions of options and trading strategies can be found in MacMillan (1980), Hull (1993), Gemmill (1992) and the opening chapters of Cox & Rubinstein (1985).
- Hull (1993) describes the workings of futures markets in some detail. Cox, Ingersoll & Ross (1981) establish the equivalence of forward and futures prices using an arbitrage argument.
- For a more mathematical treatment of many aspects of finance see Merton (1990).

Exercises

1. It is customary for shares in the UK to have prices between 100p and 1000p (in the US, between \$10 and \$100), perhaps because then typical daily changes are of the same sort of size as the last digit or two, and perhaps so that average purchase sizes for retail investors are a sensible number of shares. A company whose share price rises above this range will usually issue new shares to bring it back. This is called a **scrip issue** in the UK, a **stock split** in the US. What is the effect of a one-for-one issue (i.e. one additional, newly-created, new share for each old one) on the share price? How should option contracts be altered? What will be the effect on option prices? Illustrate with an example such as Reuters from Figure 1.1. Repeat for a two-for-one issue.

2. A stock price is S just before a dividend D is paid. What is the price immediately after the payment?

3. Should the value of call and put options increase with uncertainty? Why?

4. "If taxes and transaction costs are ignored, options transactions are a zero-sum game." What is meant by this?

2 Asset Price Random Walks

2.1 Introduction

Since the mid-1980s it has been impossible for newspaper readers or television viewers to be unaware of the nature of financial time series. The values of the major indices (*Financial Times Stock Exchange* 100, or *FT-SE*, in the UK, the *S&P 500* and *Dow Jones* in the US and the *Nikkei Dow* in Japan) are quoted frequently. Graphs of these indices appear on television news bulletins throughout the day. As an extreme example of a financial time series, Figure 2.1 shows the *FT-SE* daily closing prices for the six months each side of the October 1987 stock market crash. To many people these 'mountain ranges' showing the variation of the value of an asset[1] or index with time are excellent examples of 'random walks'.

It must be emphasised that this book is *not* about the prediction of asset prices. Indeed, our basic assumption, common to most of option pricing theory, is that we *do not know* and *cannot predict* tomorrow's values of asset prices. The past history of the asset value is there as a financial time series for us to examine as much as we want, but we cannot use it to forecast the next move that the asset will make. This does not mean that it tells us nothing. We know from our examination of the past what are the likely jumps in asset price, what are their mean and variance and, generally, what is the likely distribution of future asset prices. These qualities must be determined by a statistical analysis of historical data. Since this is not a statistical text, we assume that we

[1] We use the word 'asset' for any financial product whose value is quoted or can, in principle, be measured. Examples include equities, indices, currencies and commodities.

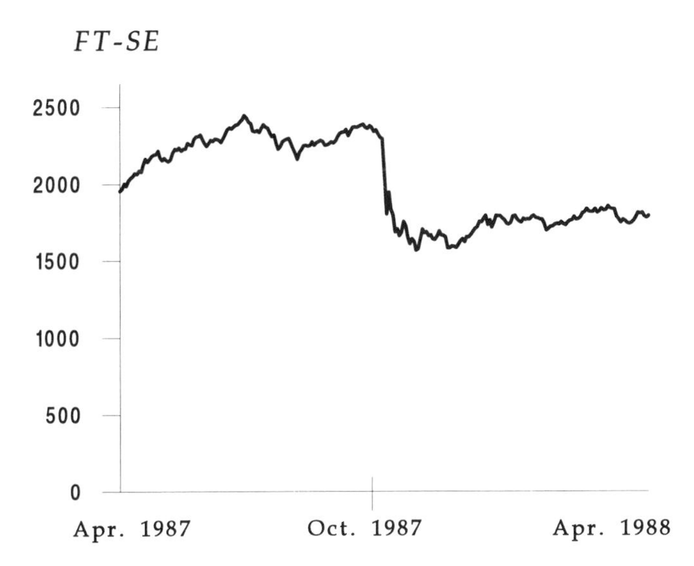

Figure 2.1. *FT-SE* closing prices from April 1987 to April 1988.

know them, although a brief discussion is given in the Technical Point at the end of the next section.

Almost all models of option pricing are founded on one simple model for asset price movements, involving parameters derived, for example, from historical or market data. This chapter is devoted to a discussion of this model.

2.2 A Simple Model for Asset Prices

It is often stated that asset prices must move randomly because of the **efficient market hypothesis**. There are several different forms of this hypothesis with different restrictive assumptions, but they all basically say two things:

- The past history is fully reflected in the present price, which does not hold any further information;
- Markets respond immediately to any new information about an asset.

Thus the modelling of asset prices is really about modelling the arrival of new information which affects the price. With the two assumptions above, unanticipated changes in the asset price are a **Markov process**.

Firstly, we note that the *absolute* change in the asset price is not by itself a useful quantity: a change of 1p is much more significant when the asset price is 20p than when it is 200p. Instead, with each change in asset price, we associate a **return**, defined to be the change in the price divided by the original value. This relative measure of the change is clearly a better indicator of its size than any absolute measure.

Now suppose that at time t the asset price is S. Let us consider a small subsequent time interval dt, during which S changes to $S + dS$, as sketched in Figure 2.2. (We use the notation $d\cdot$ for the small change in any quantity over this time interval when we intend to consider it as an infinitesimal change.) How might we model the corresponding return on the asset, dS/S? The commonest model decomposes this return into two parts. One is a predictable, deterministic and anticipated return akin to the return on money invested in a risk-free bank. It gives a contribution

$$\mu\, dt$$

to the return dS/S, where μ is a measure of the average rate of growth of the asset price, also known as the drift. In simple models μ is taken to be a constant. In more complicated models, for exchange rates, for example, μ can be a function of S and t.

The second contribution to dS/S models the random change in the asset price in response to external effects, such as unexpected news. It is represented by a random sample drawn from a normal distribution with mean zero and adds a term

$$\sigma\, dX$$

to dS/S. Here σ is a number called the **volatility**, which measures the standard deviation of the returns. The quantity dX is the sample from a normal distribution, which is discussed further below.

Putting these contributions together, we obtain the **stochastic differential equation**

$$\frac{dS}{S} = \sigma\, dX + \mu\, dt, \tag{2.1}$$

which is the mathematical representation of our simple recipe for generating asset prices.

The only symbol in (2.1) whose role is not yet entirely clear is dX. If we were to cross out the term involving dX, by taking $\sigma = 0$, we would

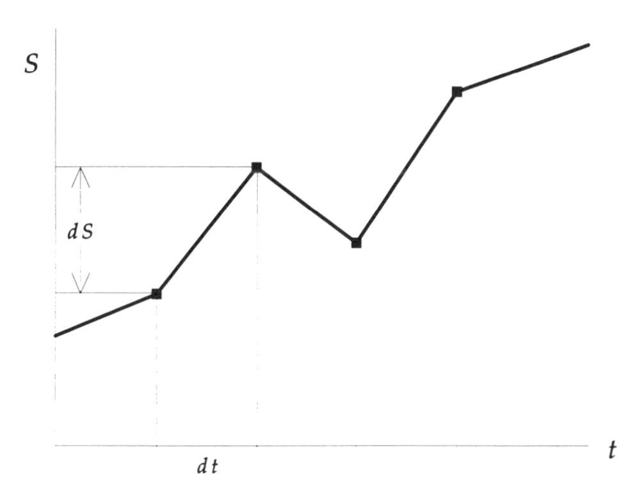

Figure 2.2. Detail of a discrete random walk.

be left with the ordinary differential equation

$$\frac{dS}{S} = \mu \, dt$$

or

$$\frac{dS}{dt} = \mu S.$$

When μ is constant this can be solved exactly to give exponential growth in the value of the asset, i.e.

$$S = S_0 e^{\mu(t-t_0)},$$

where S_0 is the value of the asset at $t = t_0$. Thus if $\sigma = 0$ the asset price is totally deterministic and we can predict the future price of the asset with certainty.

The term dX, which contains the randomness that is certainly a feature of asset prices, is known as a **Wiener process**. It has the following properties:

- dX is a random variable, drawn from a normal distribution;
- the mean of dX is zero;
- the variance of dX is dt.

One way of writing this is

$$dX = \phi\sqrt{dt},$$

where ϕ is a random variable drawn from a standardised normal distribution. The standardised normal distribution has zero mean, unit variance and a probability density function given by

$$\frac{1}{\sqrt{2\pi}}e^{-\frac{1}{2}\phi^2} \tag{2.2}$$

for $-\infty < \phi < \infty$. If we define the expectation operator $\mathcal{E}$ by

$$\mathcal{E}[F(\cdot)] = \frac{1}{\sqrt{2\pi}}\int_{-\infty}^{\infty} F(\phi)e^{-\frac{1}{2}\phi^2}\,d\phi, \tag{2.3}$$

for any function F, then

$$\mathcal{E}[\phi] = 0$$

and

$$\mathcal{E}[\phi^2] = 1.$$

The reason that dX is scaled with $\sqrt{dt}$ is that any other choice for the magnitude of dX would lead to a problem that is either meaningless or trivial when we finally consider what happens in the limit $dt \to 0$, in which we are particularly interested for the reasons given above. (We also mention that if dX were not scaled in this way, the variance of the random walk for S would have a limiting value of 0 or ∞.) We return to this point later.

We have given some economically reasonable justification for the model (2.1). A more practical justification for it is that it fits real time series data very well, at least for equities and indices. (The agreement with currencies is less good, especially in the long term.) There are some discrepancies; for instance, real data appears to have a greater probability of large rises or falls than the model predicts. But, on the whole, it has stood the test of time remarkably well and can be the starting point for more sophisticated models. As an example of such generalisation, the coefficients of dX and dt in (2.1) could be any functions of S and/or t. The particular choice of functions is a matter for the mathematical modeller and statistician, and different assets may be best represented by other stochastic differential equations.

Equation (2.1) is a particular example of a **random walk**. It cannot be solved to give a deterministic path for the share price, but it can give interesting and important information concerning the behaviour of

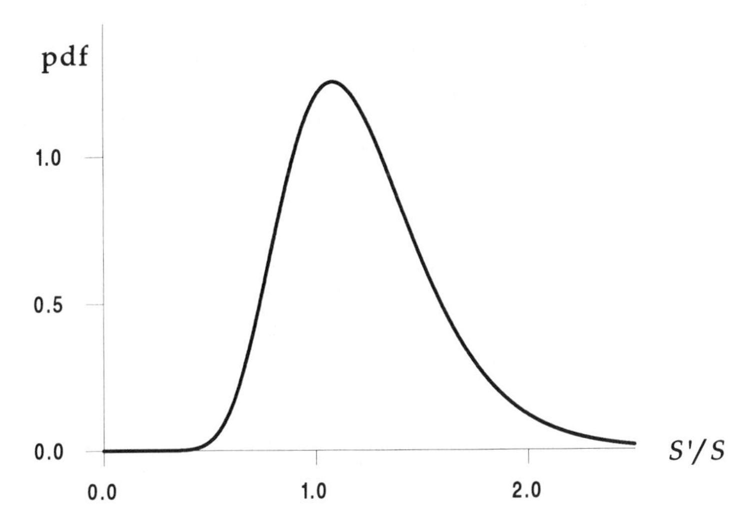

Figure 2.3. The probability density function (pdf) for S'/S.

S in a probabilistic sense. Suppose that today's date is t_0 and today's asset price is S_0. If the price at a later date t', in six months' time, say, is S', then S' will be distributed about S_0 with a probability density function of the form shown in Figure 2.3. The future asset price, S', is thus most likely to be close to S_0 and less likely to be far away. The further that t' is from t_0 the more spread out this distribution is. If S follows the random walk given by (2.1) then the probability density function represented by this skewed bell-shaped curve is the lognormal distribution (we show this below) and the random walk (2.1) is therefore known as a lognormal random walk.

We can think of (2.1) as a recipe for generating a time series – each time the series is restarted a different path results. Each path is called a **realisation** of the random walk. This recipe works as follows. Suppose, as an example, that today's price is \$1, and we have $\mu = 1$, $\sigma = 0.2$ with $dt = 1/250$ (one day as a proportion of 250 business days per year). We now draw a number at random from a normal distribution with mean zero and variance $1/250$; this is dX. Suppose that we draw the number $dX = 0.08352\ldots$. Now perform the calculation in (2.1) to find dS:

$$dS = 0.2 \times \$1.0 \times 0.08352\ldots + 1.0 \times \$1.0 \times \frac{1}{250} = \$0.020704\ldots.$$

Add this value for dS to the original value for S to arrive at the new value for S after one time-step: $S + dS = \$1.020704\ldots$. Repeat the above steps, using the new value for S and drawing a new random number. As this procedure is repeated it generates a time series of random numbers which appears similar to genuine series from the stock market, such as that in Figure 2.1.

Firstly, let us now briefly consider some of the properties of (2.1). Equation (2.1) does not refer to the past history of the asset price; the next asset price $(S + dS)$ depends solely on today's price. This independence from the past is called the **Markov property**. Secondly, we consider the mean of dS:

$$\mathcal{E}[dS] = \mathcal{E}[\sigma S\, dX + \mu S\, dt] = \mu S\, dt,$$

since $\mathcal{E}[dX] = 0$. On average, the next value for S is higher than the old by an amount $\mu S\, dt$.

Thirdly, the variance of dS is

$$\mathrm{Var}[dS] = \mathcal{E}[dS^2] - \mathcal{E}[dS]^2 = \mathcal{E}[\sigma^2 S^2 dX^2] = \sigma^2 S^2 dt.$$

The square root of the variance is the standard deviation, which is thus proportional to σ.

If we compare two random walks with different values for the parameters μ and σ, we see that the one with the larger value of μ usually rises more steeply and the one with the larger value of σ appears more jagged. Typically, for stocks and indices the value of σ is in the range 0.05 to 0.4 (the units of σ^2 are *per annum*). Government bonds are examples of assets with low volatility, while 'penny shares' and shares in high-tech companies generally have high volatility. The volatility is often quoted as a percentage, so that $\sigma = 0.2$ would be called a 20% volatility.

In the next section we learn how to manipulate functions of random variables.

Technical Point: Parameter Estimation.

None of the analysis that we have presented so far is of much use unless we can estimate the parameters in our random walk. In particular, we find later that only the volatility parameter, σ, in the random walk (2.1) appears in the value of an option. How can we estimate σ, for example from historic data?

This is not a statistics textbook; see the section on Further Reading for references on parameter estimation. A simple approach is as follows.

Suppose that we have the values of the asset price S at $n+1$ equal time-steps; closing prices, say. Call these values $S_0, \dots, S_n$ in chronological order with S_0 the first value.

Since we are assuming that changes in the asset price follow (2.1), where dX is normally distributed, we can use the usual unbiased variance estimate $\bar{\sigma}^2$ for σ^2. Let

$$\bar{m} = \frac{1}{n\,dt}\sum_{i=0}^{n-1}\frac{S_{i+1}-S_i}{S_i};$$

then

$$\bar{\sigma}^2 = \frac{1}{(n-1)\,dt}\sum_{i=0}^{n-1}\Big((S_{i+1}-S_i)/S_i - \bar{m}\,dt\Big)^2.$$

The time-step between data points, dt, is assumed to be constant, and if measured as a fraction of a year the resulting parameters are annualised.

There is a great deal more to the subject of parameter estimation, for example sizes of data sets or time dependence, but this book is not the place to discuss them.

2.3 Itô's Lemma

In real life asset prices are quoted at discrete intervals of time. There is thus a practical lower bound for the basic time-step dt of our random walk (2.1). If we used this time-step in practice to value options, though, we would find that we had to deal with unmanageably large amounts of data. Instead, we set up our mathematical models in the **continuous time** limit $dt \to 0$; it is much more efficient to solve the resulting differential equations than it is to value options by direct simulation of the random walk on a practical timescale. In order to do this, we need some technical machinery that enables us to handle the random term dX as $dt \to 0$, and this is the content of this section.

Itô's lemma is the most important result about the manipulation of random variables that we require. It is to functions of random variables what Taylor's theorem is to functions of deterministic variables, in that it relates the small change in a function of a random variable to the small change in the random variable itself. Our heuristic approach to Itô's lemma is based on the Taylor series expansion; for a more rigorous yet still readable analysis, see the books referred to at the end of the chapter.

Before coming to Itô's lemma we need one result, which we do not prove rigorously (see Technical Point 1 below). This result is that, with probability 1,

$$dX^2 \to dt \quad \text{as} \quad dt \to 0. \tag{2.4}$$

Thus, the smaller dt becomes, the more certainly dX^2 is equal to dt.

Suppose that $f(S)$ is a smooth function of S and forget for the moment that S is stochastic. If we vary S by a small amount dS then clearly f also varies by a small amount provided we are not close to singularities of f. From the Taylor series expansion we can write

$$df = \frac{df}{dS}dS + \frac{1}{2}\frac{d^2f}{dS^2}dS^2 + \cdots, \tag{2.5}$$

where the dots denote a remainder which is smaller than any of the terms we have retained. Now recall that dS is given by (2.1). Here dS is simply a number, albeit random, and so squaring it we find that

$$\begin{aligned} dS^2 &= (\sigma S\,dX + \mu S\,dt)^2 \\ &= \sigma^2 S^2 dX^2 + 2\sigma\mu S^2 dt\,dX + \mu^2 S^2 dt^2. \end{aligned} \tag{2.6}$$

We now examine the order of magnitude of each of the terms in (2.6). (See Technical Point 2 below for the symbol $O(\cdot)$.) Since

$$dX = O(\sqrt{dt}),$$

the first term is the largest for small dt and dominates the other two terms. Thus, to leading order,

$$dS^2 = \sigma^2 S^2 dX^2 + \cdots.$$

Since $dX^2 \to dt$, to leading order

$$dS^2 \to \sigma^2 S^2 dt.$$

We substitute this into (2.5) and retain only those terms which are at least as large as $O(dt)$. Using also the definition of dS from (2.1), we find that

$$\begin{aligned} df &= \frac{df}{dS}(\sigma S\,dX + \mu S\,dt) + \tfrac{1}{2}\sigma^2 S^2 \frac{d^2f}{dS^2}dt \\ &= \sigma S\frac{df}{dS}dX + \left(\mu S\frac{df}{dS} + \tfrac{1}{2}\sigma^2 S^2 \frac{d^2f}{dS^2}\right)dt. \end{aligned} \tag{2.7}$$

This is Itô's lemma[2] relating the small change in a function of a random variable to the small change in the variable itself.

Because the order of magnitude of dX is $O(\sqrt{dt})$, the second derivative of f with respect to S appears in the expression for df at order dt. The order dt terms play a significant part in our later analyses, and any other choice for the order of dX would not lead to the interesting results we discover. It can be shown that any other order of magnitude for dX leads to unrealistic properties for the random walk in the limit $dt \to 0$; if $dX \gg \sqrt{dt}$ the random variable goes immediately to zero or infinity, and if $dX \ll \sqrt{dt}$ the random component of the walk vanishes in the limit $dt \to 0$.

Observe that (2.7) is made up of a random component proportional to dX and a deterministic component proportional to dt. In this respect it bears a resemblance to equation (2.1). Equation (2.7) is also a recipe, this time for determining the behaviour of f, and f itself follows a random walk.

The result (2.7) can be further generalised by considering a function of the random variable S and of time, $f(S, t)$. This entails the use of partial derivatives since there are now *two* independent variables, S and t. We can expand $f(S + dS, t + dt)$ in a Taylor series about (S, t) to get

$$df = \frac{\partial f}{\partial S} dS + \frac{\partial f}{\partial t} dt + \frac{1}{2} \frac{\partial^2 f}{\partial S^2} dS^2 + \cdots .$$

Using our expressions (2.1) for dS and (2.4) for dX^2 we find that the new expression for df is

$$df = \sigma S \frac{\partial f}{\partial S} dX + \left(\mu S \frac{\partial f}{\partial S} + \tfrac{1}{2} \sigma^2 S^2 \frac{\partial^2 f}{\partial S^2} + \frac{\partial f}{\partial t} \right) dt. \qquad (2.8)$$

As a simple example of the theory above, consider the function $f(S) = \log S$. Differentiation of this function gives

$$\frac{df}{dS} = \frac{1}{S} \quad \text{and} \quad \frac{d^2 f}{dS^2} = -\frac{1}{S^2}.$$

[2] We have here applied Itô's lemma to functions of the random variable S, which is defined by (2.1). The lemma is, of course, more general than this and can be applied to functions of any random variable, G, say, described by a stochastic differential equation of the form

$$dG = A(G, t)\, dX + B(G, t)\, dt.$$

Thus given $f(G)$, Itô's lemma says that

$$df = A \frac{df}{dG} dX + \left(B \frac{df}{dG} + \tfrac{1}{2} A^2 \frac{d^2 f}{dG^2} \right) dt.$$

Thus, using (2.7), we arrive at

$$df = \sigma\,dX + \left(\mu - \tfrac{1}{2}\sigma^2\right) dt.$$

This is a constant coefficient stochastic differential equation, which says that the jump df is normally distributed. Now consider f itself: it is the sum of the jumps df (in the limit, the sum becomes an integral). Since a sum of normal variables is also normal, $f - f_0$ has a normal distribution with mean $(\mu - \frac{1}{2}\sigma^2)t$ and variance $\sigma^2 t$. (Here, of course, $f_0 = \log S_0$ is the initial value of f.) The probability density function of $f(S)$ is therefore

$$\frac{1}{\sigma\sqrt{2\pi t}} e^{-\left(f - f_0 - (\mu - \frac{1}{2}\sigma^2)t\right)^2/2\sigma^2 t} \tag{2.9}$$

for $-\infty < f < \infty$.

Now that we have the probability density function of $f(S) = \log S$, it is not difficult to show (the derivation is left as an exercise) that the probability density function of S itself is

$$\frac{1}{\sigma S\sqrt{2\pi t}} e^{-\left(\log(S/S_0) - (\mu - \frac{1}{2}\sigma^2)t\right)^2/2\sigma^2 t} \tag{2.10}$$

for $0 < S < \infty$; (2.10) is known as the **lognormal distribution**, and the random walk that gives rise to it is often called a lognormal random walk. We shall use it later, when we discuss risk neutrality in Chapter 5 and in the binomial method in Chapter 10.

Technical Point 1: The Limit of dX^2 as $dt \to 0$.

To be technically correct we should write the stochastic differential equation (2.1) in the *integrated* form

$$S(t) = S(t_0) + \sigma \int_{t_0}^{t} S\,dX + \mu \int_{t_0}^{t} S\,dt.$$

All the theory for stochastic calculus is based on this representation of a random walk and, strictly speaking, (2.1) is only shorthand notation.

We do not yet have a definition for the term involving the integration with respect to the Wiener process. One definition of such integrals, due to Itô, is that, for any function h,

$$\mathrm{Int} = \int_{t_0}^{t} h(\tau)\,dX(\tau) = \lim_{m\to\infty} \mathrm{Int}_m$$

where

$$\mathrm{Int}_m = \sum_{k=0}^{m-1} h(t_k)\big(X(t_{k+1}) - X(t_k)\big). \tag{2.11}$$

Here $t_0 < t_1 \ldots < t_m = t$ is any partition (or division) of the range $[t_0, t]$ into m smaller regions and X is the running sum of the random variables dX. The important point to note about (2.11) is that the value of the function h inside the summation is taken at the left-hand end of the small regions, i.e. at $t = t_k$ and not at t_{k+1}. (This is, in effect, where the Markov property is incorporated into the model.)

If $X(t)$ were a smooth function the integral would be the usual Stieltjes integral and it would not matter that h was evaluated at the left-hand end. However, because of the randomness, which does not go away as $dt \to 0$, the fact that the summation depends on the left-hand value of h in each partition becomes important. For example,

$$\int_{t_0}^{t} X(\tau)\, dX(\tau) = \tfrac{1}{2}\left(X(t)^2 - X(t_0)^2\right) - \tfrac{1}{2}(t - t_0). \tag{2.12}$$

The last term would not be present if X were smooth.

Using the formal definition of stochastic integration it can be shown that

$$f(S(t)) = f(S(t_0)) + \int_{t_0}^{t} \sigma S \frac{df}{dS}\, dX + \int_{t_0}^{t} \left(\mu S \frac{df}{dS} + \tfrac{1}{2}\sigma^2 S^2 \frac{d^2 f}{dS^2} \right) dt,$$

which when written in the shorthand notation becomes (2.7) as 'derived' above. We can conclude that the rules for differentiation and integration are different from those of classical calculus, but can generally be derived heuristically by remembering the simple rule of thumb

$$dX^2 = dt.$$

Technical Point 2: Order Notation.

Order notation is a convenient shorthand representation of the idea that some complicated quantity, such as a term in an equation, is 'about the same size as' some other, usually simpler, quantity. Suppose that $F(t)$ and $G(t)$ are two functions of t and that, as $t \to 0$,

$$F(t) \le C G(t)$$

for some constant C (equivalently, $\lim_{t\to 0} F(t)/G(t)$ is bounded by C). Then we write

$$F(t) = O\big(G(t)\big) \qquad \text{as} \quad t \to 0.$$

There is nothing special about $t = 0$ in this definition; we could have been concerned with any value of t (including infinity). If the limit of

$F(t)/G(t)$ is actually 1, it is usual to write

$$F(t) \sim G(t) \qquad \text{as} \quad t \to 0,$$

although conventions differ on the exact interpretation of the symbol $\sim$ ('twiddles'); it is sometimes taken to be equivalent to $O(\cdot)$. If $F(t)/G(t) \to 0$ as $t \to 0$, we write

$$F(t) = o\big(G(t)\big) \qquad \text{as} \quad t \to 0;$$

this is sometimes abbreviated to

$$F(t) \ll G(t).$$

In the discussion of Itô's lemma above, we have both $dX = O(\sqrt{dt})$ as $dt \to 0$ and $dX \sim \sqrt{dt}$ as $dt \to 0$. We see also that $dX\,dt = o(dt)$ as $dt \to 0$ (or $dX\,dt \ll dt$), and this is why we are able to ignore terms of this size in Itô's lemma.

2.4 The Elimination of Randomness

The two random walks in S (equation (2.1)) and f (equation (2.8)) are both driven by the single random variable dX. We can exploit this fact to construct a third variable g whose variation dg is wholly deterministic during the small time period dt. For the moment this appears to be merely a clever trick but it takes on major importance when we come to value options.

Let Δ be a number at our disposal and let

$$g = f - \Delta S$$

where Δ is held constant during the time-step dt. We can write

$$\begin{aligned} dg &= df - \Delta\, dS \\ &= \sigma S \frac{\partial f}{\partial S} dX + \left(\mu S \frac{\partial f}{\partial S} + \tfrac{1}{2}\sigma^2 S^2 \frac{\partial^2 f}{\partial S^2} + \frac{\partial f}{\partial t} \right) dt \\ &\quad -\Delta(\sigma S\, dX + \mu S\, dt) \\ &= \sigma S \left(\frac{\partial f}{\partial S} - \Delta \right) dX \\ &\quad + \left(\mu S \left(\frac{\partial f}{\partial S} - \Delta \right) + \tfrac{1}{2}\sigma^2 S^2 \frac{\partial^2 f}{\partial S^2} + \frac{\partial f}{\partial t} \right) dt. \end{aligned}$$

(If Δ were allowed to vary during the time-step then in evaluating dg we would need to include terms in $d\Delta$.) Now, by choosing $\Delta = \partial f/\partial S$

(evaluated before the jumps, i.e. at time t) we can make the coefficient of dX vanish. This leaves a value for dg which is known: the random walk for g is purely deterministic. Essentially, this 'trick' used the fact that the two random walks, for S and for f, are correlated and so not independent. Since their random components are proportional, by taking the correct linear combination of f and S the random variation (or risk) can be eliminated altogether. This is just the argument we used informally in Section 1.4, and in the next chapter it turns out to be crucial in the discussion of option pricing.

Further Reading

- Cox & Rubinstein (1985) give a good description of the **binomial model** in which asset prices do not change continuously in time but rather jump at discrete intervals to one of two new values. Such discrete models, although not necessarily accurate models of the real world, can often give insight into financial problems.
- Jump-diffusion models are discussed by Jarrow & Rudd (1983) and Merton (1976). In these models asset prices behave as we have described with one additional property: they can occasionally undergo random jumps of a substantial fraction of their value.
- For a further and more detailed description of the movement of equity prices see Brealey (1983), Fama (1965) and Mandelbrot (1963).
- See Spiegel (1980) for general details of parameter estimation, and Leong (1993) for information specific to option pricing.
- See Schuss (1980) or Øksendal (1992) for accounts of stochastic calculus.
- *Option Pricing* has more details on the probability distribution of an asset that follows a random walk, and of 'technical indicators' such as moving averages.
- Chapter 3 of Merton (1990) deals with the question of the order of magnitude of dX.

Exercises

1. If $dS = \sigma S\, dX + \mu S\, dt$, and A and n are constants, find the stochastic differential equations satisfied by

   (a) $f(S) = AS$, (b) $f(S) = S^n$.

2. Use Itô's lemma to confirm that equation (2.12) is correct.

3. Derive (2.10) from (2.9).

4. Consider the general stochastic differential equation

$$dG = A(G,t)\,dX + B(G,t)\,dt.$$

Use Itô's lemma to show that it is theoretically possible to find a function $f(G)$ which itself follows a random walk but with zero drift.

5. There are n assets satisfying the following stochastic differential equations:

$$dS_i = \sigma_i S_i\,dX_i + \mu_i S_i\,dt \quad \text{for} \quad i = 1, \ldots, n.$$

The Wiener processes dX_i satisfy

$$\mathcal{E}[dX_i] = 0, \quad \mathcal{E}[dX_i^2] = dt$$

as usual, but the asset price changes are correlated with

$$\mathcal{E}[dX_i\,dX_j] = \rho_{ij}\,dt$$

where $-1 \le \rho_{ij} = \rho_{ji} \le 1$.
Derive Itô's lemma for a function $f(S_1, \ldots, S_n)$ of the n assets $S_1, \ldots, S_n$.

3 The Black–Scholes Model

3.1 Introduction

We begin this chapter with a discussion of the concept of arbitrage, a concept which, in certain circumstances, allows us to establish precise relationships between prices and thence to determine them. We then discuss option strategies in general and use arbitrage, together with the model for asset price movements that we discussed in the previous chapter, to derive the celebrated Black–Scholes differential equation for the price of the simplest options, the so-called European vanilla options. We also discuss the boundary conditions to be satisfied by different types of option, and we set the scene for the derivation of explicit solutions. *This chapter is fundamental to the whole subject of option pricing and should be read with care.*

3.2 Arbitrage

One of the fundamental concepts underlying the theory of financial derivative pricing and hedging is that of **arbitrage**. This can be loosely stated as "there's no such thing as a free lunch." More formally, in financial terms, there are never any opportunities to make an instantaneous risk-free profit. (More correctly, such opportunities cannot exist for a significant length of time before prices move to eliminate them.) The financial application of this principle leads to some elegant modelling.

Almost all finance theory, this book included, assumes the existence of **risk-free** investments that give a guaranteed return[1] with no chance

[1] As explained in Chapter 17 the return available may depend on the time for which the deposit is made; the different rates available for different periods reflect the possibility that interest rates may change in the future. We need only assume that a known guaranteed short-term return is always available.

of default. A good approximation to such an investment is a government bond or a deposit in a sound bank. The greatest risk-free return that one can make on a portfolio of assets is the same as the return if the equivalent amount of cash were placed in a bank.

The key words in the definition of arbitrage are 'instantaneous' and 'risk-free'; by investing in equities, say, one can *probably* beat the bank, but this cannot be *certain*. If one wants a greater return then one must accept a greater risk. Why should this be so? Suppose that an opportunity did exist to make a guaranteed return of greater magnitude than from a bank deposit. Suppose also that most investors behave sensibly. Would any sensible investor put money in the bank when putting it in the alternative investment yields a greater return? Obviously not. Moreover, if he could also borrow money at less than the return on the alternative investment then he should borrow as much as possible from the bank to invest in the higher-yielding opportunity. In response to the pressure of supply and demand we would expect the bank to raise its interest rates to attract money and/or the yield from the other investment to drop. There is some elasticity in this argument because of the presence of 'friction' factors such as transaction costs, differences in borrowing and lending rates, problems with liquidity, tax laws, etc., but on the whole the principle is sound since the market place *is* inhabited by **arbitragers** whose (highly paid) job it is to seek out and exploit irregularities or mispricings such as the one we have just illustrated.

Technical Point: Risk.
Risk is commonly described as being of two types: specific and non-specific. (The latter is also called market or systematic risk.) Specific risk is the component of risk associated with a single asset (or a sector of the market, for example chemicals), whereas non-specific risk is associated with factors affecting the whole market. An unstable management would affect an individual company but not the market; this company would show signs of specific risk, a highly volatile share price perhaps. On the other hand the possibility of a change in interest rates would be a non-specific risk, as such a change would affect the market as a whole.

It is often important to distinguish between these two types of risk because of their behaviour within a large portfolio (a **portfolio** is a term for a collection of investments). Provided one has a sensible definition of risk, it is possible to diversify away specific risk by having a portfolio with a large number of assets from different sectors of the market; however, it is not possible to diversify away non-specific risk. (Market risk can be eliminated from a portfolio by taking similar positions in two assets which are highly negatively correlated – as one increases in value the other decreases. This is not diversification but hedging, which is of the utmost importance in the analysis of derivatives.) It is commonly said that specific risk is not rewarded, and that only the taking of greater non-specific risk should be rewarded by a greater return.

A popular definition of the risk of a portfolio is the variance of the return. A bank account which has a guaranteed return, at least in the short term, has no variance and is thus termed riskless or risk-free. On the other hand, a highly volatile stock with a very uncertain return and thus a large variance is a risky asset. This is the simplest and commonest definition of risk, but it does not take into account the distribution of the return, but rather only one of its properties, the variance. Thus as much weight is attached to the possibility of a greater than expected return as to the possibility of a less than expected return. Other, more sophisticated, definitions of risk avoid this property and attach different weights to different returns.

3.3 Option Values, Payoffs and Strategies

Now we turn to option pricing. Let us introduce some simple notation, which we use consistently throughout the book.

- We denote by V the value of an option; when the distinction is important we use $C(S,t)$ and $P(S,t)$ to denote a call and a put respectively. This value is a function of the current value of the underlying asset, S, and time, t: $V = V(S,t)$. The value of the option also depends on the following parameters:
- σ, the volatility of the underlying asset;
- E, the exercise price;
- T, the expiry;
- r, the interest rate.

First, consider what happens just at the moment of expiry of a call option, that is, at time $t = T$. A simple arbitrage argument tells us its value at this special time.

If $S > E$ at expiry, it makes financial sense to exercise the call option, handing over an amount E, to obtain an asset worth S. The profit from such a transaction is then $S - E$. On the other hand, if $S < E$ at expiry, we should not exercise the option because we would make a loss of $E - S$.

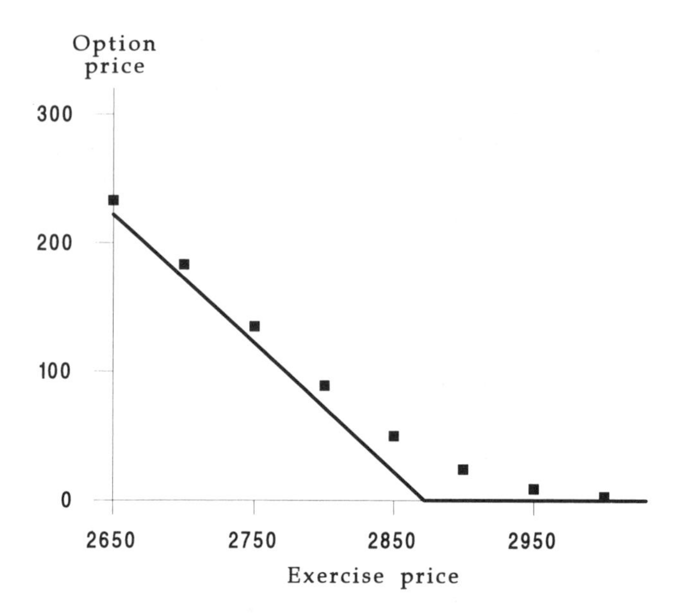

Figure 3.1 The value of a call option at and before expiry against exercise price; option values from *FT-SE* index option data.

In this case, the option expires worthless. Thus, the value of the call option at expiry can be written as

$$C(S,T) = \max(S - E, 0). \tag{3.1}$$

As we get nearer to the expiry date we can expect the value of our call option to approach (3.1). To confirm this we reproduce in Figure 3.1 the figure from Chapter 1 which compares real *FT-SE* index call option data with the value of the option at expiry for fixed S. In this figure we show $\max(S - E, 0)$ as a function of E for fixed S ($= 2872$) and superimpose the real data for V taken from the February call option series. Observe that the real data is always just above the predicted line. This reflects the fact that there is still some time remaining before the option expires – there is potential left for the asset price to rise further, giving the option even greater value. This difference between the option

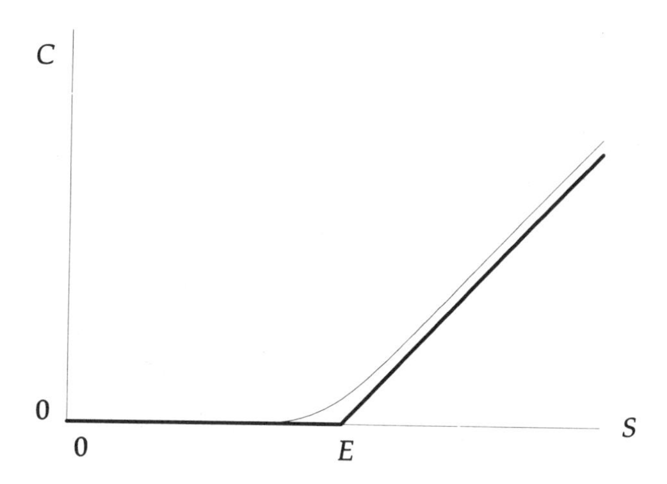

Figure 3.2 The payoff diagram for a call, $C(S,T)$, and the option value, $C(S,t)$, prior to expiry, as functions of S.

value before and at expiry is called the **time value** and the value at expiry the **intrinsic value**.[2]

If one owns an option with a given exercise price, then one is less interested in how the option value varies with exercise price than with how it varies with asset price, S. In Figure 3.2 we plot

$$\max(S-E,0)$$

as a function of S (the bold line) and also the value of an option at some time before expiry. The latter curve is just a sketch of a plausible form for the option value. For the moment the reader must trust that the value of the option before expiry is of this form. Later in this chapter we see how to derive equations and sometimes formulæ for such curves.

The bold line, being the payoff for the option at expiry, is called a **payoff diagram**. The reader should be aware that some authors use the term 'payoff diagram' or 'profit diagram' to mean the difference between the terminal value of the contract (*our* payoff) and the original premium. We choose not to use this definition for two reasons. Firstly,

[2] Other important jargon is **at-the-money**, which refers to that option whose exercise price is closest to the current value of the underlying asset, **in-the-money**, which is a call (put) whose exercise price is less (greater) than the current asset price – so that the option value has a significant intrinsic component – and **out-of-the-money**, which is a call or put with no intrinsic value.

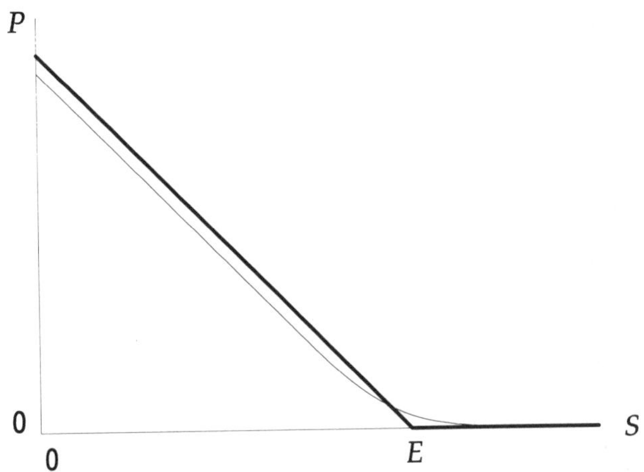

Figure 3.3 The payoff diagram for a put, $P(S,T)$, and the option value, $P(S,t)$, prior to expiry, as functions of S.

the premium is paid at the start of the option contract and the return, if any, only comes at expiry. Secondly, the payoff diagram has a natural interpretation, as we see, as the final condition for a diffusion equation.

It should now be clear that each option and portfolio of options has its own payoff at expiry. An argument similar to that given above for the value of a call at expiry leads to the payoff for a put option. At expiry it is worthless if $S > E$ but has the value $E - S$ for $S < E$. Thus the payoff at expiry for a put option is

$$\max(E - S, 0).$$

The payoff diagram for a European put is shown in Figure 3.3, where the bold line shows the payoff function $\max(E - S, 0)$. The other curve is again a sketch of the option value prior to expiry. Although the time value of the call option of Figure 3.2 is everywhere positive, for the put the time value is negative for sufficiently small S, where the option value falls below the payoff. We return to this point later.

Although the two most basic structures for the payoff are the call and the put, in principle there is no reason why an option contract cannot be written with a more general payoff. An example of another payoff is shown in Figure 3.4. This payoff can be written as

$$B\mathcal{H}(S - E),$$

where $\mathcal{H}(\cdot)$ is the **Heaviside function**, which has value 0 when its argument is negative but is 1 otherwise. This option may be interpreted

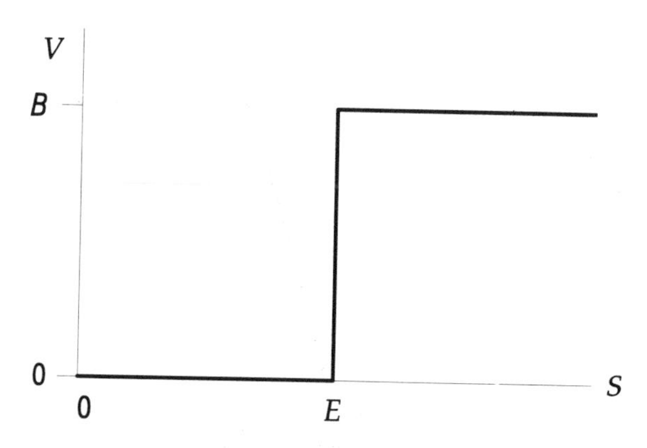

Figure 3.4 The payoff diagram for a cash-or-nothing call, equivalent to a bet on the asset price.

as a straight bet on the asset price; it is called a **cash-or-nothing call.** Options with general payoffs are usually called **binaries** or **digitals.**

By combining calls and puts with various exercise prices one can construct portfolios with a great variety of payoffs. For example, we show in Figure 3.5 the payoff for a 'bullish vertical spread', which is constructed by buying one call option and writing one call option with the same expiry date but a larger exercise price. This portfolio is called 'bullish' because the investor profits from a rise in the asset price, 'vertical' because there are two different exercise prices involved, and 'spread' because it is made up of the same type of option, here calls. The payoff function for this portfolio can be written as

$$\max(S - E_1, 0) - \max(S - E_2, 0)$$

with $E_2 > E_1$.

Many other portfolios can be constructed. Some examples are 'combinations', containing both calls and puts, and 'horizontal' or 'calendar' spreads, containing options with different expiry dates. Others are given in the exercises at the end of this chapter.

The appeal of such strategies is in their ability to redirect risk. In exchange for the premium – which is the maximum possible loss and known from the start – one can construct portfolios to benefit from virtually any move in the underlying asset. If one has a view on the market and this turns out to be correct then, as we have seen, one can

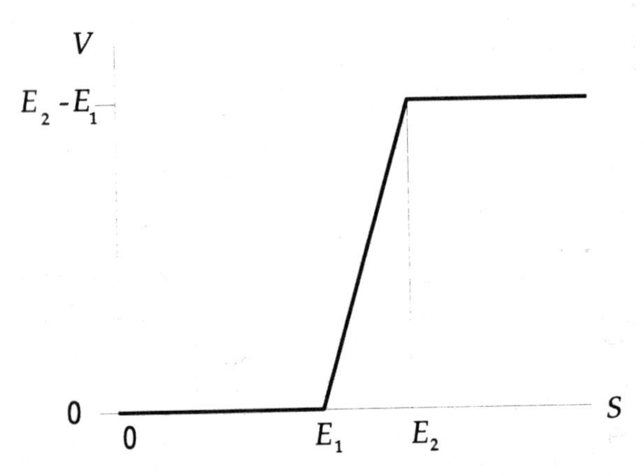

Figure 3.5. The payoff diagram for a bullish vertical spread.

make large profits from relatively small movements in the underlying asset.

3.4 Put-call Parity

Although call and put options are superficially different, in fact they can be combined in such a way that they are perfectly correlated. This is demonstrated by the following argument.

Suppose that we are long one asset, long one put and short one call. The call and the put both have the same expiry date, T, and the same exercise price, E. Denote by Π the value of this portfolio. We thus have

$$\Pi = S + P - C,$$

where P and C are the values of the put and the call respectively. The payoff for this portfolio at expiry is

$$S + \max(E - S, 0) - \max(S - E, 0).$$

This can be rewritten as

$$S + (E - S) - 0 = E \quad \text{if} \quad S \leq E,$$

or

$$S + 0 - (S - E) = E \quad \text{if} \quad S \geq E.$$

Whether S is greater or less than E at expiry the payoff is always the same, namely E.

Now ask the question

- How much would I pay for a portfolio that gives a guaranteed E at $t = T$?

This is, of course, the same question that we asked in Chapter 1, and the answer is arrived at by discounting the final value of the portfolio. (Note that here we do have to assume the existence of a known risk-free interest rate over the lifetime of the option.) Thus this portfolio is now worth $Ee^{-r(T-t)}$. This equates the return from the portfolio with the return from a bank deposit. If this were not the case then arbitragers could (and would) make an instantaneous riskless profit: by buying and selling options and shares and at the same time borrowing or lending money in the correct proportions, they could lock in a profit today with zero payoff in the future. Thus we conclude that

$$S + P - C = Ee^{-r(T-t)}. \tag{3.2}$$

This relationship between the underlying asset and its options is called **put-call parity**. It is an example of risk elimination, achieved by carrying out one transaction in the asset and each of the options. In the next section, we see that a more sophisticated version of this idea, involving a continuous rebalancing, rather than the one-off transactions above, allows us to value European call and put options independently.

3.5 The Black–Scholes Analysis

Before describing the Black–Scholes analysis which leads to the value of an option we list the assumptions that we make for most of the book.

- The asset price follows the lognormal random walk (2.1).

  Other models do exist, and in many cases it is possible to perform the Black–Scholes analysis to derive a differential equation for the value of an option. Explicit formulæ rarely exist for such models. However, this should not discourage their use, since an accurate numerical solution is usually quite straightforward.
- The risk-free interest rate r and the asset volatility σ are known functions of time over the life of the option.

  Only in Chapters 17 and 18 do we drop the assumption of deterministic behaviour of r; there we model interest rates by a stochastic differential equation.

- There are no transaction costs associated with hedging a portfolio.
  In Chapter 16 we describe a model which allows for transaction costs.
- The underlying asset pays no dividends during the life of the option.
  This assumption can be dropped if the dividends are known beforehand. They can be paid either at discrete intervals or continuously over the life of the option. We discuss this point further in Chapter 6.
- There are no arbitrage possibilities.
  The absence of arbitrage opportunities means that all risk-free portfolios must earn the same return.
- Trading of the underlying asset can take place continuously.
  This is clearly an idealisation, and becomes important in the chapter on transaction costs, Chapter 16.
- Short selling is permitted and the assets are divisible.
  We assume that we can buy and sell any number (not necessarily an integer) of the underlying asset, and that we may sell assets that we do not own.

Suppose that we have an option whose value $V(S,t)$ depends only on S and t. It is not necessary at this stage to specify whether V is a call or a put; indeed, V can be the value of a whole portfolio of different options although for simplicity the reader can think of a simple call or put. Using Itô's lemma, equation (2.8), we can write

$$dV = \sigma S \frac{\partial V}{\partial S} dX + \left(\mu S \frac{\partial V}{\partial S} + \tfrac{1}{2}\sigma^2 S^2 \frac{\partial^2 V}{\partial S^2} + \frac{\partial V}{\partial t} \right) dt. \qquad (3.3)$$

This gives the random walk followed by V. Note that we require V to have at least one t derivative and two S derivatives.

Now construct a portfolio consisting of one option and a number $-\Delta$ of the underlying asset. This number is as yet unspecified. The value of this portfolio is

$$\Pi = V - \Delta S. \qquad (3.4)$$

The jump in the value of this portfolio in one time-step is

$$d\Pi = dV - \Delta\, dS.$$

Here Δ is held fixed during the time-step; if it were not then $d\Pi$ would contain terms in $d\Delta$. Putting (2.1), (3.3) and (3.4) together, we find that Π follows the random walk

$$d\Pi = \sigma S\left(\frac{\partial V}{\partial S} - \Delta\right)dX + \left(\mu S\frac{\partial V}{\partial S} + \tfrac{1}{2}\sigma^2 S^2\frac{\partial^2 V}{\partial S^2} + \frac{\partial V}{\partial t} - \mu\Delta S\right)dt. \tag{3.5}$$

As we demonstrated in Section 2.4, we can eliminate the random component in this random walk by choosing

$$\Delta = \frac{\partial V}{\partial S}. \tag{3.6}$$

Note that Δ is the value of $\partial V/\partial S$ at the *start* of the time-step dt.

This results in a portfolio whose increment is wholly deterministic:

$$d\Pi = \left(\frac{\partial V}{\partial t} + \tfrac{1}{2}\sigma^2 S^2\frac{\partial^2 V}{\partial S^2}\right)dt. \tag{3.7}$$

We now appeal to the concepts of arbitrage and supply and demand, with the assumption of no transaction costs. The return on an amount Π invested in riskless assets would see a growth of $r\Pi\,dt$ in a time dt. If the right-hand side of (3.7) were greater than this amount, an arbitrager could make a guaranteed riskless profit by borrowing an amount Π to invest in the portfolio. The return for this risk-free strategy would be greater than the cost of borrowing. Conversely, if the right-hand side of (3.7) were less than $r\Pi\,dt$ then the arbitrager would short the portfolio and invest Π in the bank. Either way the arbitrager would make a riskless, no cost, instantaneous profit. The existence of such arbitragers with the ability to trade at low cost ensures that the return on the portfolio and on the riskless account are more or less equal. Thus, we have

$$r\Pi\,dt = \left(\frac{\partial V}{\partial t} + \tfrac{1}{2}\sigma^2 S^2\frac{\partial^2 V}{\partial S^2}\right)dt. \tag{3.8}$$

Substituting (3.4) and (3.6) into (3.8) and dividing throughout by dt we arrive at

$$\frac{\partial V}{\partial t} + \tfrac{1}{2}\sigma^2 S^2\frac{\partial^2 V}{\partial S^2} + rS\frac{\partial V}{\partial S} - rV = 0. \tag{3.9}$$

This is the **Black–Scholes partial differential equation**. With its extensions and variants, it plays the major role in the rest of the book.

It is hard to overemphasise the fact that, under the assumptions stated earlier, *any* derivative security whose price depends *only* on the current value of S and on t, and which is paid for up-front, must satisfy the Black–Scholes equation (or a variant incorporating dividends or time-dependent parameters). Many seemingly complicated option valuation problems, such as exotic options, become simple when looked at in this way. It is also important to note, though, that many options, for example American options, have values that depend on the history of the asset price as well as its present value. We see later how they fit into the Black–Scholes framework.

Before moving on, we make three remarks about the derivation we have just seen. Firstly, the **delta**, given by

$$\Delta = \frac{\partial V}{\partial S},$$

is the rate of change of the value of our option or portfolio of options with respect to S. It is of fundamental importance in both theory and practice, and we return to it repeatedly. It is a measure of the correlation between the movements of the option or other derivative products and those of the underlying asset.

Secondly, the linear differential operator $\mathcal{L}_{BS}$ given by

$$\mathcal{L}_{BS} = \frac{\partial}{\partial t} + \tfrac{1}{2}\sigma^2 S^2 \frac{\partial^2}{\partial S^2} + rS\frac{\partial}{\partial S} - r$$

has a financial interpretation as a measure of the difference between the return on a hedged option portfolio (the first two terms) and the return on a bank deposit (the last two terms). Although this difference must be identically zero for a European option, in order to avoid arbitrage, we see later that this need not be so for an American option.

Thirdly, we note that the Black–Scholes equation (3.9) does not contain the growth parameter μ. In other words, the value of an option is independent of how rapidly or slowly an asset grows. The only parameter from the stochastic differential equation (2.1) for the asset price that affects the option price is the volatility, σ. A consequence of this is that two people may differ in their estimates for μ yet still agree on the value of an option.

3.6 The Black–Scholes Equation

Equation (3.9) is the first partial differential equation that we have derived in this book. The theory and solution methods for partial differential equations are discussed in depth in Chapters 4 and 5; nevertheless, we now introduce a few basic points in the theory so that the reader is aware of what we are trying to achieve.

By deriving the partial differential equation for a quantity, such as an option price, we have made an enormous step towards finding its value.

We hope to be able to find an expression for this value by solving the equation. Sometimes this involves solution by numerical means if exact formulæ cannot be found. However, a partial differential equation on its own generally has many solutions; for example, the values of puts, calls and S itself all satisfy the Black–Scholes equation. The value of an option should be unique (otherwise, arbitrage possibilities would arise), and so, to pin down the solution, we must also impose boundary conditions. A boundary condition specifies the behaviour of the required solution at some part of the solution domain.

The most frequent type of partial differential equation in financial problems is the parabolic equation. A parabolic equation for a function $V(S,t)$ is a specific relationship between V and its partial derivatives with respect to the independent variables S and t. In the simplest case, the highest derivative with respect to S is a second derivative, and the highest derivative with respect to t is only a first derivative. Thus (3.9) comes into this category. If the equation is linear and the signs of these particular derivatives are the same, when they appear on the same side of the equation, then the equation is called backward parabolic; otherwise it is called forward parabolic. Equation (3.9) is backward parabolic.

Once we have decided that our partial differential equation is of this parabolic type we can make general statements about the sort of boundary conditions that lead to a unique solution. Typically, we must pose two conditions in S, which has the second derivative associated with it, but only one in t. For example we could specify that

$$V(S,t) = V_a(t) \quad \text{on} \quad S = a$$

and

$$V(S,t) = V_b(t) \quad \text{on} \quad S = b$$

where V_a and V_b are given functions of t.

If the equation is of backward type we must also impose a 'final' condition such as

$$V(S,t) = V_T(S) \quad \text{on} \quad t = T$$

where V_T is a known function. We then solve for V in the region $t < T$. That is, we solve 'backwards in time', hence the name. If the equation is of forward type we impose an 'initial' condition on $t = 0$, say, and solve in $t > 0$, in the forward direction. Of course, we can change from backward to forward by the simple change of variables $t' = -t$. This is why both types of equation are mathematically equivalent and it is common to transform backward equations into forward equations before any analysis. It is important to remember, however, that the parabolic equation cannot be solved in the wrong direction; that is, we should not impose initial conditions on a backward equation.

3.7 Boundary and Final Conditions for European Options

Having derived the Black–Scholes equation for the value of an option, we must next consider final and boundary conditions, for otherwise the partial differential equation does not have a unique solution. For the moment we restrict our attention to a European call, with value now denoted by $C(S,t)$, with exercise price E and expiry date T.

The final condition, to be applied at $t = T$, comes from the arbitrage argument described in Section 3.3. At $t = T$, the value of a call is known with certainty to be the payoff:

$$C(S,T) = \max(S - E, 0). \tag{3.10}$$

This is the final condition for our partial differential equation.

Our 'spatial' or asset-price boundary conditions are applied at zero asset price, $S = 0$, and as $S \to \infty$. We can see from (2.1) that if S is ever zero then dS is also zero and therefore S can never change. This is the only deterministic case of the stochastic differential equation (2.1). If $S = 0$ at expiry the payoff is zero. Thus the call option is worthless on $S = 0$ even if there is a long time to expiry. Hence on $S = 0$ we have

$$C(0,t) = 0. \tag{3.11}$$

As the asset price increases without bound it becomes ever more likely that the option will be exercised and the magnitude of the exercise price becomes less and less important. Thus as $S \to \infty$ the value of the option becomes that of the asset and we write

$$C(S,t) \sim S \quad \text{as} \quad S \to \infty. \tag{3.12}$$

For a European call option, without the possibility of early exercise, (3.9)–(3.12) can be solved exactly to give the Black–Scholes value of a call option. We show how to do this in Chapter 5, and at the end of this section we quote the results for a European call and put.

For a put option, with value $P(S,t)$, the final condition is the payoff

$$P(S,T) = \max(E - S, 0). \tag{3.13}$$

We have already mentioned that if S is ever zero then it must remain zero. In this case the final payoff for a put is known with certainty to be E. To determine $P(0,t)$ we simply have to calculate the present value of an amount E received at time T. Assuming that interest rates are constant we find the boundary condition at $S=0$ to be

$$P(0,t) = Ee^{-r(T-t)}. \tag{3.14}$$

More generally, for a time-dependent interest rate we have

$$P(0,t) = Ee^{-\int_t^T r(\tau)\,d\tau}.$$

As $S \to \infty$ the option is unlikely to be exercised and so

$$P(S,t) \to 0 \quad \text{as} \quad S \to \infty. \tag{3.15}$$

Technical Point: Boundary Conditions at Infinity.
We see later that we can transform (3.9) into an equation with constant coefficients by the change of variable $S = Ee^x$. The point $S=0$ becomes $x=-\infty$ and $S=\infty$ becomes $x=\infty$. As we also see, a physical analogy to the financial problem is the flow of heat in an *infinite* bar. Clearly, prescribing the temperature of the bar at $x=\pm\infty$ has no effect whatsoever at finite values of x unless the temperature is highly singular there. If the temperature at infinity is well-behaved then the temperature in any finite region of the bar is governed wholly by the initial data: it cannot be influenced by the ends at infinity. Since most option problems can be transformed into the diffusion equation it is also not strictly necessary to prescribe the boundary conditions at $S=0$ and $S=\infty$. We only need to insist that the value of the option is not too singular.

We can distinguish between

- prescribing a boundary condition in order to make the solution unique, and
- determining the solution in the neighbourhood of the boundary, perhaps to assist or check a numerical solution.

The boundary conditions (3.11) and (3.12) contain more information than is strictly mathematically *necessary* (see Section 4.3.2). Nevertheless, they are financially *useful:* they tell us more information about the behaviour of the option at certain special parts of the S-axis and can be used to improve the accuracy of any numerical method. It can be shown that an even more accurate expression for the behaviour of C as $S \to \infty$ is

$$C(S,t) \sim S - Ee^{-r(T-t)}. \tag{3.16}$$

This is a simple correction to (3.12) which accounts for the discounted exercise price.

Throughout the book we give boundary conditions to show the local behaviour of the option price.

3.8 The Black–Scholes Formulæ for European Options

Here we quote the exact solution of the European call option problem (3.9)–(3.12) when the interest rate and volatility are constant; in Chapter 5 we show how to derive it systematically. In Chapter 6 we drop the constraint that r and σ are constant and find more general formulæ.

When r and σ are constant the exact, explicit solution for the European call is

$$C(S,t) = SN(d_1) - Ee^{-r(T-t)}N(d_2), \qquad (3.17)$$

where $N(\cdot)$ is the cumulative distribution function for a standardised normal random variable, given by

$$N(x) = \frac{1}{\sqrt{2\pi}} \int_{-\infty}^{x} e^{-\frac{1}{2}y^2}\, dy.$$

Here

$$d_1 = \frac{\log(S/E) + (r + \frac{1}{2}\sigma^2)(T-t)}{\sigma\sqrt{T-t}}$$

and

$$d_2 = \frac{\log(S/E) + (r - \frac{1}{2}\sigma^2)(T-t)}{\sigma\sqrt{T-t}}.$$

For a put, i.e. (3.9), (3.13), (3.14) and (3.15), the solution is

$$P(S,t) = Ee^{-r(T-t)}N(-d_2) - SN(-d_1). \qquad (3.18)$$

It is easy to show that these satisfy put-call parity (3.2).

The delta for a European call is

$$\Delta = \frac{\partial C}{\partial S} = N(d_1), \qquad (3.19)$$

and for a put it is

$$\Delta = \frac{\partial P}{\partial S} = N(d_1) - 1.$$

The latter follows from the former by put-call parity.

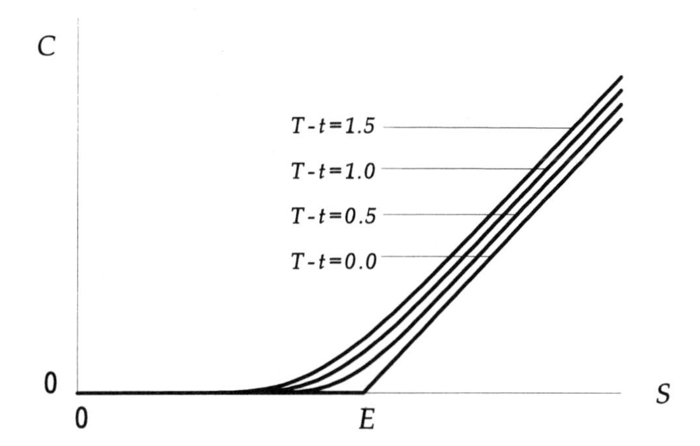

Figure 3.6 The European call value $C(S,t)$ as a function of S for several values of time to expiry; $r = 0.1$, $\sigma = 0.2$, $E = 1$ and $T - t = 0$, 0.5, 1.0 and 1.5.

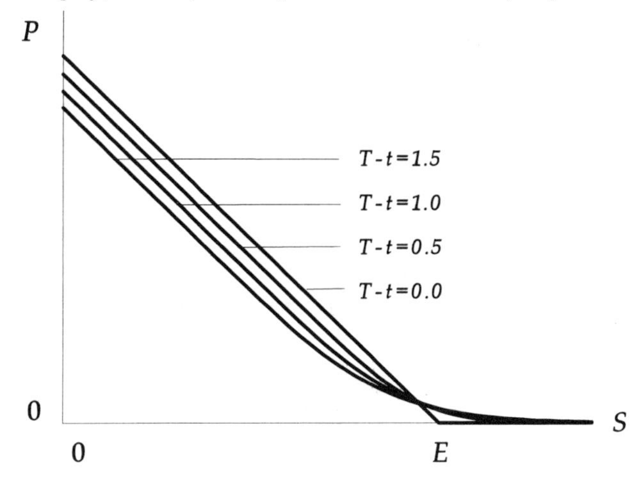

Figure 3.7 The European put value $P(S,t)$ as a function of S for several values of time to expiry; $r = 0.1$, $\sigma = 0.2$, $E = 1$ and $T - t = 0$, 0.5, 1.0 and 1.5.

Other derivatives of the option value (with respect to S, t, r and σ) can play important roles in hedging and are discussed briefly at the end of this chapter.

In Figures 3.6 and 3.7 we show plots of the European call and put values for several times up to expiry. Note how the curves approach the payoff functions as $t \to T$. In Figure 3.8 we show the European call delta as a function of S, again for several times up to expiry. The delta is always between zero and one, and approaches a step function

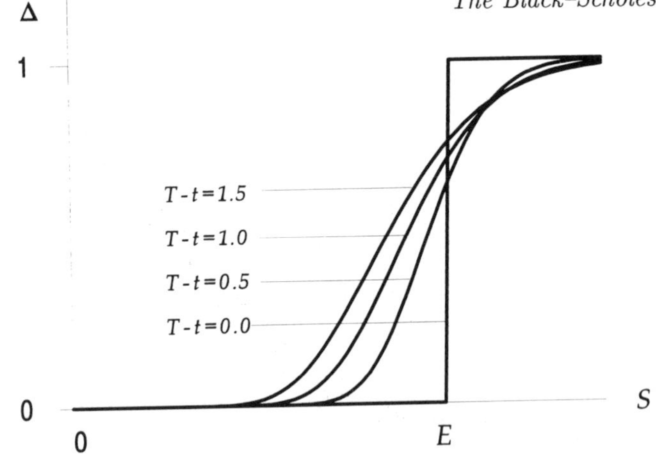

Figure 3.8 The European call delta as a function of S for several values of time to expiry; $r = 0.1$, $\sigma = 0.2$, $E = 1$ and $T - t = 0$, 0.5, 1.0 and 1.5.

as $t \rightarrow T$. Recall that the writer of a call option will be required to deliver the asset if $S > E$ at expiry, and not otherwise. If he follows the delta-hedging strategy, with a portfolio $\Delta S - C$, he will automatically hold the correct amount (one or zero units) of the asset at expiry. This is to be expected, since delta-hedging is a risk-free strategy right up to expiry. If the option expires in-the-money, the required asset will have been bought over the lifetime of the option, firstly in setting up the initial hedge, and secondly in a series of transactions as S changes. The cost of these purchases and/or sales, less the exercise price E, is exactly balanced by the initial premium and bank interest. Conversely, if the option expires out-of-the-money, the initial hedge is gradually sold. (It should also be noted that if the values of the asset just before expiry are close to E, the hedge may change from nearly zero to nearly one many times. This is awkward to handle in practice, since each transaction incurs costs. We discuss transaction costs further in Chapter 16.)

Equations (3.17) and (3.18) for the values of European call and put options are interesting in that they contain the function for the cumulative normal distribution $N(x)$. Thus the value of an option is related to the probability density function for the random variable $\log S$. It can be shown, and we discuss this in Chapter 5, that the value of an option has a natural interpretation as a certain discounted expected value of the payoff at expiry. This leads to the subject of the 'risk-neutral valuation' of contingent claims, a phrase which is explained there.

3.9 Hedging in Practice

Hedging is the reduction of the sensitivity of a portfolio to the movement of an underlying asset by taking opposite positions in different financial instruments. Two extreme cases have been introduced above; in both cases the sensitivity of the portfolio was reduced to zero. The first example was in the demonstration of put-call parity for European options and the second was in the Black–Scholes analysis with delta-hedging. These are, however, fundamentally different hedging strategies. The former involves a one-off transaction in three products (a call, a put and the underlying); the resulting portfolio can then be left unattended with the riskless return locked in. The latter is a dynamic strategy; the delta hedge is only instantaneously risk-free, and it requires a continuous rebalancing of the portfolio and the ratio of the holdings in the asset and the derivative product. The delta-hedged position must be monitored continually, and in practice it can suffer from losses due to the costs of transacting in the underlying.

One use for delta-hedging is for the writer of an option who also wishes to cover his position. If the writer can get a premium slightly above the fair value for the option then he can trade in the underlying (or a futures contract on the underlying, since this is usually cheaper to trade in because the transaction costs are lower) to maintain a delta-neutral position until expiry. Since he charges more for the option than it was theoretically worth he makes a net profit without any risk – in theory. This is only a practical policy for those with access to the markets at low dealing costs, such as market makers. If the transaction costs are significant then the frequent rehedging necessary to maintain a delta-neutral position renders the policy impractical. We discuss this point further in Chapter 16.

The delta for a whole portfolio is the rate of change of the value of the portfolio with respect to changes in the underlying asset.[3] Writing Π for the value of the portfolio,

$$\Delta = \frac{\partial \Pi}{\partial S}.$$

Thus, when delta hedging between an option and an asset, the position taken is called 'delta-neutral', since the sensitivity of the hedged portfolio to asset price changes is instantaneously zero. For a general portfolio

[3] This definition, which is standard, is not quite consistent with our previous use of Δ and Π (which is also standard). There, Δ was the sensitivity of a single option to asset price changes and Π was the hedged portfolio. There should be little risk of confusion.

the maintenance of a delta-neutral position may require a short position in the underlying asset. This entails the selling of assets which are not owned – so-called short selling. A broker may require a margin to cover any movements against the short seller but this margin usually receives interest at the bank rate.

There are more sophisticated trading strategies than simple delta-hedging, and here we mention only the basics. In delta-hedging the largest random component of the portfolio is eliminated. One can be more subtle and hedge away smaller order effects due, for instance, to the curvature (the second derivative) of the portfolio value with respect to the underlying asset. This entails knowledge of the **gamma** of a portfolio, defined by

$$\Gamma = \frac{\partial^2 \Pi}{\partial S^2}.$$

The decay of time value in a portfolio is represented by the **theta**, given by

$$\Theta = -\frac{\partial \Pi}{\partial t}.$$

Finally, sensitivity to volatility is usually called the **vega** and is given by

$$\frac{\partial \Pi}{\partial \sigma},$$

and sensitivity to interest rate is called **rho**, where

$$\rho = \frac{\partial \Pi}{\partial r}.$$

Hedging against any of these dependencies requires the use of another option as well as the asset itself. With a suitable balance of the underlying asset and other derivatives, hedgers can eliminate the short-term dependence of the portfolio on movements in time, asset price, volatility or interest rate.

3.10 Implied Volatility

We have suggested in the above modelling and analysis that the way to use the Black–Scholes and other models is to take parameter values estimated from historical data, substitute them into a formula (or perhaps solve an equation numerically), and so derive the value for a derivative product. This is no longer the commonest use of option models, at least not for the simplest options. This is partly because of difficulty in measuring the value of the volatility of the underlying asset. Despite our assumption to the contrary, it does not appear to be the case that volatility is constant for long periods of time. Furthermore, it is not obvious that the historic volatility is independent of the time series from which it is calculated, nor that it accurately predicts the future volatility that we require, over the lifetime of an option.

A direct measurement of volatility is therefore difficult in practice. However, despite these difficulties it is plainly true that option prices are quoted in the market. This suggests that, even if we do not know the volatility, the market 'knows' it. Take the Black–Scholes formula for a call, for example, and substitute in the interest rate, the price of the underlying, the exercise price and the time to expiry. All of these are very simple to measure and are either quoted constantly or are defined as part of the option contract. All that remains is to specify the volatility and the option price follows. Since a call option price increases monotonically with volatility (this is easy to show from the explicit formula and, as we have already mentioned, is clear financially) there is a one-to-one correspondence between the volatility and the option price. Thus we could take the option price quoted in the market and, working backwards, deduce the market's opinion of the value for the volatility over the remaining life of the option. This volatility, derived from the quoted price for a single option, is called the **implied volatility**.

There are more advanced ways of calculating the market view of volatility using more than one option price. In particular, using option prices for a variety of expiry dates one can, in principle, deduce the market's opinion of the future values for the volatility of the underlying (the **term structure of volatility**).

One unusual feature of implied volatility is that the implied volatility does not appear to be constant across exercise prices. That is, if the value of the underlying, the interest rate and the time to expiry are fixed, the prices of options across exercise prices should reflect a uniform value for the volatility. In practice this is not the case and this highlights a flaw in some part of the model. (Also, puts and calls tend to give slightly different implied volatilities.) Which part of the model is inaccurate is the subject of a great deal of academic research. We illustrate this effect in Figure 3.9, which shows the implied volatilities as a function of exercise price using the *FT-SE* index option data in Figure 1.1. Observe how the volatility of the options deeply in-the-money is greater than for those at-the-money. This curve is traditionally called the 'smile', although depending on market conditions it may be lopsided as in Figure 3.9, or even a 'frown'.

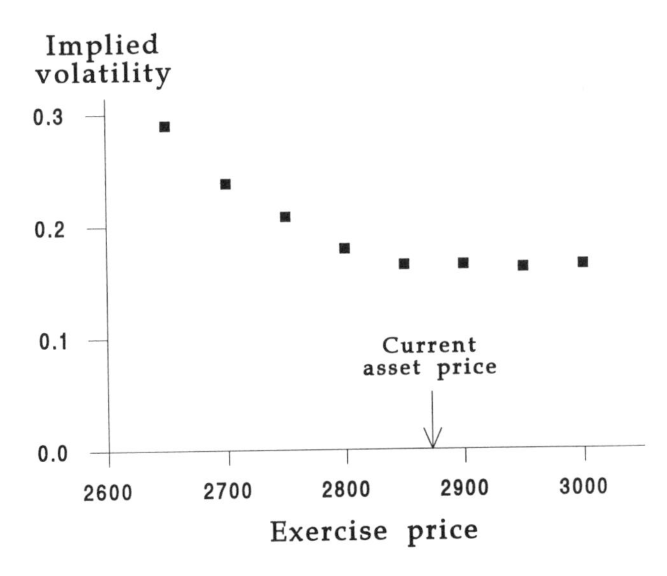

Figure 3.9 Implied volatilities as a function of exercise price. Data is taken from *FT-SE* index option prices.

Technical Point: Trading Volatility.
In practice volatility is not constant, nor is it predictable for timescales of more than a few months. This, of course, limits the validity of any model that assumes the contrary. This problem may be reduced by pricing options using implied volatility as described above. Thus one trading strategy is to calculate implied volatilities from prices of all options on the same underlying and the same expiry date and then to buy the one with the lowest volatility and sell the one with the highest. The hope is then that the prices move so that implied volatilities become more or less comparable and the portfolio makes a profit.

More sophisticated modelling involves describing volatility itself as a random variable satisfying some stochastic differential equation. This results in a **two-factor model**. If the volatility is random then it is no longer possible to construct the perfect hedge, in which a portfolio grows by a deterministic amount, using the asset alone. However, it is in principle possible to use other options, but the details are too complex to go into here.

Further Reading

- Carefully read the original papers of Black & Scholes (1973) and Merton (1973).
- Compare the binomial method for valuing options with the differential equation approach. The binomial method can be found in, for example, Cox & Rubinstein (1985). We discuss it in Chapter 10.
- Jarrow & Rudd (1983) and Cox & Rubinstein (1985) describe 'jump-diffusion' models and 'constant elasticity of variance' models. In the former the asset price random walk need not be continuous but can have random discontinuous jumps; in the latter the volatility can be a function of S.
- Hull (1993) considers the estimation of volatility using the implied volatilities of several options. Hull & White (1987) discuss the variation of volatility with time.
- There has been a great deal of work done on testing the validity of the Black–Scholes formulæ in practice; see Hull (1993). For details of how the call option formula stands up in practice see MacBeth & Merville (1979) and for a test of put-call parity see Klemkosky & Resnick (1979).
- Gemmill (1992) gives a practical example illustrating the practical shortcomings of the purely theoretical approach to hedging.
- More sophisticated hedging strategies are described in Cox & Rubinstein (1985).

Exercises

1. Today's date is 9 January 2000 and XYZ's share price stands at \$10. On 8 November 2000 there is to be a Presidential election and you believe that, depending on who is elected, XYZ's share price will either rise or fall by approximately 10%. Construct a portfolio of options which will do well if you are correct. Calls and puts are available with expiry dates in March, June, September, December and with strike prices of \$10 plus or minus 50¢. Draw the payoff diagram and describe the payoff mathematically.

2. Draw the expiry payoff diagrams for each of the following portfolios:
   (a) Short one share, long two calls with exercise price E (this combination is called a **straddle**);
   (b) Long one call and one put, both with exercise price E (this is also a straddle: why?);

(c) Long one call and two puts, all with exercise price E (a **strip**);
(d) Long one put and two calls, all with exercise price E (a **strap**);
(e) Long one call with exercise price E_1 and one put with exercise price E_2. Compare the three cases $E_1 > E_2$ (known as a **strangle**), $E_1 = E_2$ and $E_1 < E_2$.
(f) As (e) but also short one call and one put with exercise price E (when $E_1 < E < E_2$, this is called a **butterfly spread**).

Use the market data of Figure 1.1 to calculate the cost of an example of each portfolio. What view about the market does each strategy express?

3. Show by substitution that two exact solutions of the Black–Scholes equation (3.9) are

   (a) $V(S,t) = AS$,
   (b) $V(S,t) = Ae^{rt}$,

   where A is an arbitrary constant. What do these solutions represent and what is the Δ in each case?

4. Show that the formulæ (3.17) for a call and (3.18) for a put also satisfy (3.9) with the relevant boundary conditions (one at each of $S = 0$ and $S = \infty$) and final conditions at $t = T$. Show also that they satisfy put-call parity.

5. Sketch the graphs of the Δ for the European call and put. Suppose that the asset price now is $S = E$ (each of these options is at-the-money). Convince yourself that it is plausible that the delta-hedging strategy is self-financing for each option, in the two cases that the option expires in-the-money and out-of-the-money; look at the contract from the point of view of the writer.

6. Find the most general solution of the Black–Scholes equation that has the special form

   (a) $V = V(S)$;
   (b) $V = A(t)B(S)$.

   These are examples of 'similarity solutions', which are discussed further in Chapter 5. Time-independent options as in (a) are called **perpetual** options.

7. Use arbitrage arguments to prove the following simple bounds on European call options on an asset that pays no dividends:

   (a) $C \leq S$;
   (b) $C \geq S - Ee^{-r(T-t)}$;

(c) If two otherwise identical calls have exercise prices E_1 and E_2 with $E_1 < E_2$, then

$$0 \le C(S,t;E_1) - C(S,t;E_2) \le E_2 - E_1;$$

(d) If two otherwise identical call options have expiry times T_1 and T_2 with $T_1 < T_2$, then

$$C(S,t;T_1) \le C(S,t;T_2).$$

Derive similar restrictions for put options.

8. Derive equation (3.16).

9. Suppose that a share price S is currently \$100, and that tomorrow it will be either \$101, with probability p, or \$99, with probability $1-p$. A call option, with value C, has exercise price \$100. Set up a Black–Scholes hedged portfolio and hence find the value of C. (Ignore interest rates.)

   Now repeat the calculation for a cash-or-nothing call option with payoff \$100 if the final asset price is above \$100, zero otherwise. What difference do you notice?

   This very simple discrete model is the basis of the **binomial method**, described in Chapter 10.

4 Partial Differential Equations

4.1 Introduction

The modelling of Chapter 3 culminates in the formulation of the pricing problem for a derivative product as a partial differential equation. We now take a break from the financial modelling to discuss, in this and the next chapter, some of the theory behind such differential equations. In this chapter we describe the elementary theory and the nature of boundary and initial conditions. In Chapter 5 we derive some explicit solutions, including the original Black–Scholes formulæ. Later, in Chapter 7, we describe in detail the special problems arising when there are free boundaries. This chapter is of particular importance when considering the valuation of American options.

The study of partial differential equations in complete generality is a vast undertaking. Fortunately, however, almost all the partial differential equations encountered in financial applications belong to a much more manageable subset of the whole: second order linear parabolic equations. These technical terms are discussed below; more detailed treatments of the areas beyond the scope of this text are given in some of the references at the end of the chapter.

We begin this chapter with a review of second order linear parabolic equations: their physical interpretation, mathematical properties of their solutions, and techniques for obtaining explicit solutions to specific problems. Then, we exploit this knowledge in the context of financial models to derive explicit solutions to some option valuation problems, and we set the scene for the numerical methods of Chapters 8 and 9.

Before doing this, though, it is helpful to step back and consider in general terms the questions we should ask when considering a partial differential equation. Such questions usually include any or all of the following:

- Does the equation make sense mathematically? If it is to be solved in a region, what must we say about the solution on the boundary of that region in order to obtain a **well-posed problem**, i.e. one whose solution exists, is unique, and is, in some sense, 'well-behaved'? Such specifications of the solution on the boundary are called **boundary conditions** or, if applied at a particular value of time, **initial conditions** or **final conditions**. The term 'well-behaved' used here is usually taken to imply that the solution depends continuously on the initial and boundary conditions, so that small changes in these conditions cannot induce large changes in the solution itself. Beyond this, we also want to know what mathematical properties the solution must or can have. For example, is it guaranteed to be smooth or can it have discontinuities?
- Can we develop analytical tools to solve the equation? Explicit solutions are useful both to illustrate the general behaviour of the equation and for their application in practice. We note, though, that many explicit solutions may be so cumbersome as to be of less practical use than a well-designed numerical approximation.
- How should we solve the equation numerically, should this be necessary? What implications do the mathematical properties of the solution have for the numerical method we choose? Are there alternative formulations, such as a change of variable or a weak statement of the problem (see Chapter 7), that lead to a better (simpler, more adaptable, more accurate, more robust, faster) numerical scheme?

These aims guide us in the sections to follow.

4.2 The Diffusion Equation

The **heat** or **diffusion** equation[1]

$$\frac{\partial u}{\partial \tau} = \frac{\partial^2 u}{\partial x^2} \tag{4.1}$$

has been studied for nearly two centuries as a model of the flow (or diffusion) of heat in a continuous medium. It is one of the most successful

[1] We use x rather than S as the spatial independent variable because all our applications of the diffusion equation occur after a change of variable of the form $S = Ee^x$. We use τ as the 'time' variable rather than t for a similar reason; the details are given later.

and widely used models of applied mathematics, and a considerable body of theory on its properties and solution is available. It is often helpful as a guide to intuition to bear in mind the physical situations that lead to the heat equation, and we mention them wherever it is appropriate. Thus, we recall that equation (4.1) models the diffusion of heat in one space dimension, where $u(x,\tau)$ represents the temperature in a long, thin, uniform bar of material whose sides are perfectly insulated so that its temperature varies only with distance x along the bar and, of course, with time τ.

We begin with a list of some of the elementary properties of the diffusion equation.

- It is a **linear** equation. That is, if u_1 and u_2 are solutions, then so is $c_1u_1 + c_2u_2$ for any constants c_1 and c_2.
- It is a **second order** equation, since the highest order derivative occurring is the second, in the term $\partial^2 u/\partial x^2$.
- It is a **parabolic** equation. Its characteristics are given by $\tau =$ constant. (The terms 'parabolic' and 'characteristic' are discussed further in Technical Point 1 at the end of this section.) Thus, information propagates along these lines in (x,τ) space, and if a change is made to u at a particular point, for example on the boundary of the solution region, its effect is felt instantaneously everywhere else.
- Generally speaking, its solutions are **analytic** functions of x. This means that for each value of τ greater than the initial time, $u(x,\tau)$ regarded as a function of x has a convergent power series in terms of $x - x_0$ for each x_0 away from spatial boundaries. For practical purposes, for $\tau > 0$ we can think of a solution of the diffusion equation as being as smooth a function of x as we could ever need, but discontinuities in time may be induced by the boundary conditions. This is again a consequence of the fact that information propagates with infinite speed along the characteristics $\tau =$ constant.

From the physical point of view, diffusion is a smoothing out process: heat flows from hot to cold and so evens out temperature differences. The properties above go some way towards showing that solutions of the diffusion equation, which is a mathematical model of the physical process, have the same tendency. Anticipating some results from Section 4.3, it can be shown further that even though the initial values of u may be rather irregular or jagged, for any $\tau > 0$ the solution of the

initial value problem

$$\frac{\partial u}{\partial \tau} = \frac{\partial^2 u}{\partial x^2}, \qquad -\infty < x < \infty,$$

with initial data

$$u(x, 0) = u_0(x)$$

and

$$u \to 0 \quad \text{as} \quad x \to \pm\infty$$

is analytic for all $\tau > 0$. This smoothness, which is characteristic of all (forward) linear parabolic equations, is very helpful when it comes to numerical solution.

An illustration of all these points is the following special solution, which is derived in Section 5.2:

$$u_\delta(x, \tau) = \frac{1}{2\sqrt{\pi\tau}} e^{-x^2/4\tau} \qquad -\infty < x < \infty, \quad \tau > 0. \tag{4.2}$$

For $\tau > 0$ this is a smooth Gaussian curve, but at $\tau = 0$ it is 'equal' to the delta function (hence our notation):

$$u_\delta(x, 0) = \delta(x).$$

At $\tau = 0$, $u_\delta(x, 0)$ vanishes for $x \neq 0$; at $x = 0$ it is 'infinite', but its integral is still 1. (This is to be interpreted as follows: since for all $\tau > 0$, $\int_{-\infty}^{\infty} u_\delta(x, \tau)\, dx = 1$, the limit as τ tends to zero from above of the integral is still 1. Further information on delta functions is given in Technical Point 2 below.) We show $u_\delta(x, \tau)$ in Figure 4.1 for several values of τ; note how the curve becomes taller and narrower as τ gets smaller.

The delta function initial value for $u_\delta(x, \tau)$ says that all the heat is initially concentrated at $x = 0$. This function models the evolution of an idealised 'hotspot', a unit amount of heat initially concentrated into a single point, and it is called the **fundamental solution** of the diffusion equation. It also illustrates the infinite propagation speed mentioned above. At $\tau = 0$, the solution (4.2) is zero for all $x \neq 0$, but for any $\tau > 0$, however small, and any x, however large, $u_\delta(x, \tau) > 0$: the heat initially concentrated at $x = 0$ immediately diffuses out to all values of x. Note, though, that u_δ falls off very rapidly as $|x| \to \infty$.

Finally, note that the right-hand side of equation (4.2) is just the normal distribution of probability theory, with mean zero and variance 2τ. This solution of the diffusion equation can be interpreted as the

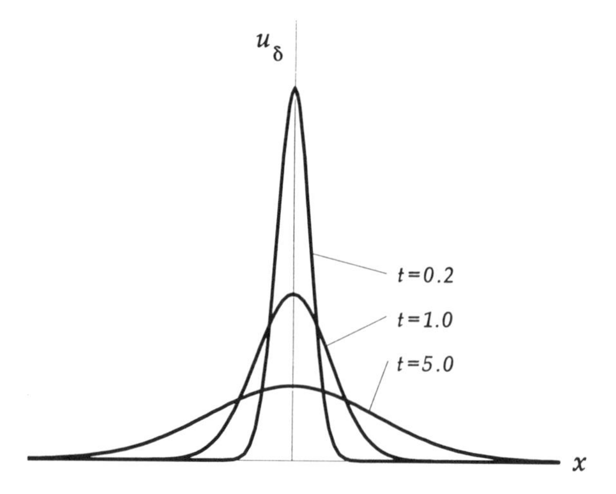

Figure 4.1. The fundamental solution of the diffusion equation.

probability density function of the future position of a particle that follows a constant coefficient random walk along the x-axis. The delta function initial condition simply says that the particle is initially known to be at the origin.

Technical Point 1: Characteristics of Second Order Linear Partial Differential Equations.

We can think of the characteristics of a second order linear equation as curves *along* which information can propagate, or as curves *across* which discontinuities in the second derivatives of u can occur. Suppose that $u(x, \tau)$ satisfies the general second order linear equation

$$a(x,\tau)\frac{\partial^2 u}{\partial x^2} + b(x,\tau)\frac{\partial^2 u}{\partial x \partial \tau} + c(x,\tau)\frac{\partial^2 u}{\partial \tau^2} + d(x,\tau)\frac{\partial u}{\partial x} + e(x,\tau)\frac{\partial u}{\partial \tau} + f(x,\tau)u + g(x,\tau) = 0.$$

The idea is to see whether the derivative terms can be written in terms of directional derivatives, so that the equation is partly like an ordinary differential equation along curves with these vectors as tangents. These curves are the characteristics. If we write them as $x = x(\xi)$, $\tau = \tau(\xi)$, where ξ is a parameter along the curves, then $x(\xi)$ and $\tau(\xi)$ satisfy

$$a(x,\tau)\left(\frac{d\tau}{d\xi}\right)^2 - b(x,\tau)\frac{d\tau}{d\xi}\frac{dx}{d\xi} + c(x,\tau)\left(\frac{dx}{d\xi}\right)^2 = 0.$$

There now arises the question whether this equation, regarded as a quadratic in $(dx/d\xi)/(d\tau/d\xi)$, has two distinct real roots, two equal real roots, or no real roots at all. These cases correspond to the discriminant b^2-4ac being greater than zero, zero, or less than zero. The first case, two real families of characteristics, is called **hyperbolic**, and is typical of wave-propagation problems. These do not often occur in finance. The second case, an exact square, is called **parabolic**; the diffusion equation, which has $b=c=0$, is the simplest example. All the second order equations in this book are parabolic. The final case, with no real characteristics, is called **elliptic**, and is typical of steady-state problems such as perpetual options in multi-factor models which are beyond the scope of this book.

Note that the definitions given here are pointwise: the hyperbolic/parabolic/elliptic distinction is specified at each point. It is possible for an equation to change type as $a(x,\tau)$, $b(x,\tau)$ and $c(x,\tau)$ vary, if the discriminant changes sign. In particular, the Black–Scholes equation (in S and t rather than x and τ),

$$\frac{\partial V}{\partial t}+\tfrac{1}{2}\sigma^2 S^2\frac{\partial^2 V}{\partial S^2}+rS\frac{\partial V}{\partial S}-rV=0,$$

is parabolic for $S>0$ (it is in fact hyperbolic at $S=0$, where it reduces to an ordinary differential equation with characteristic $S=0$). This fact has important financial implications: the line $S=0$ is a barrier across which information cannot cross.

Technical Point 2: The Delta Function and the Heaviside Function.

The Dirac delta function, written $\delta(x)$, is not in fact a function in the normal sense of the word, but is rather a 'generalised function'. For technical reasons, its definition is as a linear map, but it is really motivated by the need for a mathematical description of the limit of a function whose effect is confined to a smaller and smaller interval, but yet remains finite.

Suppose, for example, that I receive money at the rate $f(t)\,dt$ in a time dt where f is equal to the following function:

$$f(t)=\begin{cases}1/2\epsilon, & |t|\le\epsilon\\ 0, & |t|>\epsilon.\end{cases}$$

This function is drawn in Figure 4.2 for several values of ϵ. As ϵ gets smaller the graph becomes taller and narrower. It is clear that the total payment is

$$\int_{-\infty}^{\infty} f(t)\,dt$$

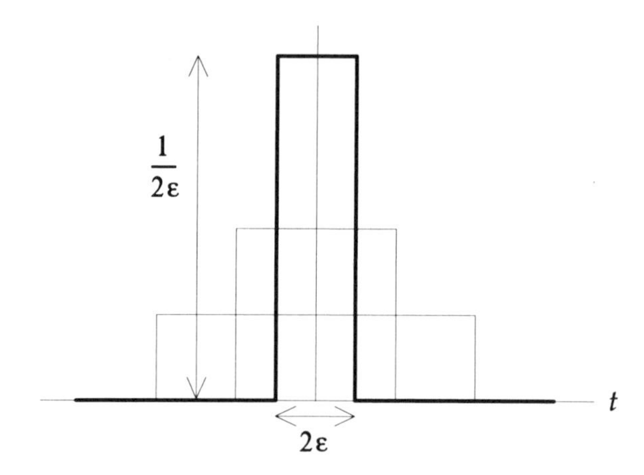

Figure 4.2. Three members of a limiting sequence for the delta function.

and is equal to 1 independently of ϵ, but that for all $t \neq 0$, $f(t) \to 0$ as $\epsilon \to 0$. The limiting 'function' is zero for all nonzero t, yet its integral is still 1! This is an informal way of defining the **delta function**, $\delta(t)$: it is the 'limit' as $\epsilon \to 0$ of any one-parameter family of functions $\delta_\epsilon(t)$ with the following properties:

- for each ϵ, $\delta_\epsilon(t)$ is piecewise smooth;
- $\displaystyle\int_{-\infty}^{\infty} \delta_\epsilon(t)\, dt = 1;$
- for each $t \neq 0$, $\quad \displaystyle\lim_{\epsilon \to 0} \delta_\epsilon(t) = 0.$

Such a sequence of functions is called a delta sequence. The function $f(t)$ above is one such; another, which uses x as the independent variable[2] instead of t, is

$$\delta_\epsilon(x) = \frac{1}{2\sqrt{\pi\epsilon}} e^{-x^2/4\epsilon}.$$

With ϵ replaced by t, this is the fundamental solution of the diffusion equation discussed above. It is easily confirmed that the latter function has integral 1, and that, like $f(t)$, for $x \neq 0$ it tends to zero as $\epsilon \to 0$, while for $x = 0$ its value increases without limit.

This 'pointwise' view of the delta function is rather hard to work with, since the functions δ_ϵ become increasingly badly behaved near the origin as $\epsilon \to 0$. Indeed, the limiting 'function' is not a normal function at all (this is why the term 'generalised function' is used). Instead, we exploit

[2] Whether we use x, t or any other letter as the argument for the delta function depends on the application we have in mind.

the fact that integration smooths out the bad behaviour; the integral of *any* member of a delta sequence *is* well-behaved, being equal to 1. This idea motivates the definition of the delta function via its integral action: for any smooth function $\phi(x)$, called a **test function**,

$$\begin{aligned}\int_{-\infty}^{\infty} \delta(x)\phi(x)\,dx &= \lim_{\epsilon\to 0}\int_{-\infty}^{\infty} \delta_\epsilon(x)\phi(x)\,dx \\ &= \phi(0).\end{aligned}$$

(In fact, this defines the delta function as the continuous linear map from smooth functions $\phi(x)$ to real numbers that has the value $\phi(0)$, usually written as $\langle\delta,\phi\rangle = \phi(0)$.)

It is apparent that for any $a, b > 0$,

$$\int_{-a}^{b} \delta(x)\phi(x)\,dx = \phi(0),$$

and that for any x_0,

$$\int_{-\infty}^{\infty} \delta(x-x_0)\phi(x)\,dx = \phi(x_0),$$

so multiplying ϕ by $\delta(x-x_0)$ and integrating 'picks out' the value of ϕ at x_0. We also have

$$\int_{-\infty}^{x} \delta(s)\,ds = \mathcal{H}(x),$$

where $\mathcal{H}(x)$ is the **Heaviside function**, defined by

$$\mathcal{H}(x) = \begin{cases} 0 & \text{for } x < 0 \\ 1 & \text{for } x \geq 0. \end{cases}$$

Conversely,

$$\mathcal{H}'(x) = \delta(x).$$

The last pair of relations shows that the derivative of a function that has a jump discontinuity has a delta function component at the same point, multiplied by the magnitude of the jump. This fact is often useful in the analysis of differential equations with discontinuous functions or coefficients. We give a simple example.

Suppose that $M(t)$ represents the amount of money owned by a person who is initially penniless, but at time $t = t_0 > 0$ receives an amount D_δ. Then, clearly,

$$M(t) = \begin{cases} 0 & \text{for } 0 < t < t_0 \\ 0 + D_\delta & \text{for } t \geq t_0, \end{cases}$$

and so $M(t)$ satisfies the differential equation

$$\frac{dM}{dt} = D_\delta\, \delta(t - t_0).$$

The discontinuity in $M(t)$ gives a delta function in dM/dt at $t = t_0$. Conversely, when we see a differential equation with a delta function on the right-hand side, there must be a corresponding delta function in the highest order derivative on the left-hand side in order to maintain a balance. This in turn means that the next highest order derivative has a jump discontinuity of magnitude equal to the coefficient of the delta function. These jump conditions can be used to join together smooth segments of the solution across discontinuities. The delta function in examples like this can be multiplied by a smooth function of x or t, but care must be taken to avoid products like $\delta(x)\mathcal{H}(x)$ or $(\delta(x))^2$, for which no sensible definition can easily be given.

4.3 Initial and Boundary Conditions

We now consider what initial and boundary conditions are appropriate for solutions of the diffusion equation, first in a finite region, then in an infinite one.

4.3.1 The Initial Value Problem on a Finite Interval

Suppose we wish to solve $\partial u/\partial \tau = \partial^2 u/\partial x^2$ in the finite interval $-L < x < L$ and for $\tau > 0$, representing heat flow in a bar of finite length $2L$.

Obviously we should specify the initial temperature $u(x, 0) = u_0(x)$ for $-L < x < L$. With the heat flow analogy in mind, it seems reasonable on physical grounds that we have enough information to determine $u(x, \tau)$ uniquely if we specify either the temperatures at the ends of the bar or the heat fluxes at both ends, but not both. This turns out to be the case; in fact both the following statements of the problem can be shown to be well-posed:

$$\text{(i)}\quad \frac{\partial u}{\partial \tau} = \frac{\partial^2 u}{\partial x^2}, \qquad -L < x < L, \text{ with } u(x, 0) = u_0(x),$$
$$u(-L, \tau) = g_-(\tau),\ u(L, \tau) = g_+(\tau);$$

(ii) $\dfrac{\partial u}{\partial \tau} = \dfrac{\partial^2 u}{\partial x^2}, \quad -L < x < L, \text{ with } u(x,0) = u_0(x),$

$$-\frac{\partial u}{\partial x}(-L,\tau) = h_-(\tau), \ \frac{\partial u}{\partial x}(L,\tau) = h_+(\tau).$$

In the first case it is the temperature and in the second case the heat fluxes that are specified at $x = -L$ and $x = L$.

4.3.2 The Initial Value Problem on an Infinite Interval

Suppose now that we consider heat flow in a very long bar, by taking the limit $L \to \infty$ in the example above. When the bar is infinitely long, it is still important to say how u behaves at large distances, but we do not have to be as precise in our specification of u at the 'boundaries' $x = \pm\infty$ as we were in the finite case. There are some technical difficulties here, associated with the notion of infinity, but roughly speaking as long as u is not allowed to grow too fast, the solution exists, is unique, and depends continuously on the initial data $u_0(x)$. To be specific, the solution to the initial value problem

$$\frac{\partial u}{\partial \tau} = \frac{\partial^2 u}{\partial x^2} \qquad -\infty < x < \infty, \quad \tau > 0, \tag{4.3}$$

with

$$u(x,0) = u_0(x), \tag{4.4}$$

where

$$\text{(i)} \qquad u_0(x) \text{ is sufficiently well-behaved,} \tag{4.5}$$

$$\text{(ii)} \qquad \lim_{|x|\to\infty} u_0(x)e^{-ax^2} = 0 \quad \text{for any } a > 0, \tag{4.6}$$

and lastly where

$$\lim_{|x|\to\infty} u(x,\tau)e^{-ax^2} = 0 \qquad \text{for any } a > 0, \ \tau > 0, \tag{4.7}$$

is well-posed. The precise definition of the phrase 'sufficiently well-behaved' here is beyond the scope of this book, but certainly any function that has no worse than a finite number of jump discontinuities is acceptable. We also note that although it is necessary to prescribe the

behaviour at infinity, in practice the limitations above are not too severe. All the initial value problems in this book satisfy the growth conditions quite comfortably.

We sometimes need to consider initial value problems defined on a semi-infinite interval, for example in the analysis of barrier options. In this case we require a combination of the two sets of conditions above. If, for example, we need to solve (4.3) for $0 < x < \infty$, $\tau > 0$, then given sufficiently smooth initial data $u_0(x)$ for $0 < x < \infty$, a sufficiently smooth boundary value at $x = 0$, $u(0, \tau) = g_0(\tau)$, and the growth conditions (4.6), (4.7) as $x \to \infty$, the problem is well-posed.

4.4 Forward versus Backward

In all the above we have discussed the **forward** equation

$$\frac{\partial u}{\partial \tau} = \frac{\partial^2 u}{\partial x^2},$$

with conditions given at $\tau = 0$. The reader may ask, what is wrong with the equation

$$\frac{\partial u}{\partial \tau} = -\frac{\partial^2 u}{\partial x^2} \tag{4.8}$$

(with the same initial and boundary conditions)? This equation might, for example, arise if in a forward problem we had replaced τ by $\tau_0 - \tau$ for some constant τ_0, whereupon $\partial u/\partial \tau$ becomes $-\partial u/\partial \tau$. It turns out that this backward problem is **ill-posed**: for most initial and boundary data the solution does not exist at all, and even if it does exist, it is likely to blow up (for example, u may tend to ∞) within a finite time. A good example is the fundamental solution of the diffusion equation (4.2). At time τ_0 this solution is equal to

$$\frac{1}{2\sqrt{\pi\tau_0}} e^{-x^2/4\tau_0},$$

which is as smooth and well-behaved as we could wish. If we use this function as our initial data $u_0(x)$ for equation (4.8), then the solution is

$$u(x, \tau) = \frac{1}{2\sqrt{\pi(\tau_0 - \tau)}} e^{-x^2/4(\tau_0 - \tau)},$$

and this becomes singular (blows up) at $\tau = \tau_0$, when it is equal to the delta function $\delta(x)$. Moreover, it cannot be continued beyond this time (at least, not as a 'normal' function).

Physically this distinction makes good sense. If the forward diffusion equation models the evolution of the temperature from its initial values, the backward equation poses the question of determining the temperature from which the initial distribution could have evolved; this is clear from the time-reversal argument above. Since forward diffusion smooths out jagged temperature distributions, backward diffusion makes smooth initial data become more jagged. Another way of seeing this is to note that under forward diffusion heat flows from hot to cold, whereas under backward diffusion it flows from cold to hot, and so the hot places become ever hotter, leading to blow-up.

There are, however, some well-posed problems for equation (4.8); in particular the **final value problem** for the backward diffusion equation is well-posed. Thus, we can solve (4.8) for $0 < \tau < \tau_0$ with $u(x, \tau_0)$ given. This is easily shown by converting (4.8) to a forward problem by replacing τ by $\tau_0 - \tau$.

Further Reading

- For further information about first order partial differential equations and their solution see Williams (1980), Strang (1986), Keener (1988) and Kevorkian (1990).
- Three books devoted wholly to the diffusion equation are those by Crank (1989), Hill & Dewynne (1990) and Carslaw & Jaeger (1989).
- More details about delta functions, and about other generalised functions (or 'distributions') are given by Richards & Youn (1990).

Exercises

1. Show that the solution to the initial value problem is unique provided that it is sufficiently smooth and decays sufficiently fast at infinity, as follows:

   Suppose that $u_1(x, \tau)$ and $u_2(x, \tau)$ are both solutions to the initial value problem (4.3)–(4.7). Show that $v(x, \tau) = u_1 - u_2$ is also a solution of (4.3) with $v(x, 0) = 0$.

   Show that if

$$E(\tau) = \int_{-\infty}^{\infty} v^2 \, dx,$$

   then

$$E(\tau) \geq 0, \qquad E(0) = 0,$$

and, by integrating by parts, that

$$\frac{dE}{d\tau} \leq 0;$$

thus $E(\tau) \equiv 0$, hence $v(x, \tau) \equiv 0$.

Note, though, that as yet we have no guarantee that $u(x, \tau)$ exists, nor that the above manipulations can be justified.

2. Show that $\sin nx\ e^{-n^2\tau}$ is a solution of the forward diffusion equation, and that $\sin nx\ e^{n^2\tau}$ is a solution of the backward diffusion equation. Now try to solve the initial value problem for the forward and backward equations in the interval $-\pi < x < \pi$, with $u = 0$ on the boundaries and $u(x, 0)$ given, by expanding the solution in a Fourier series in x with coefficients depending on τ. What difference do you see between the two problems? Which is well-posed? (The former is useful for double knockout options; see Exercise 3 in Chapter 12.)

3. Verify that $u_\delta(x, \tau)$ does satisfy the diffusion equation for $\tau > 0$.

5 The Black–Scholes Formulæ

5.1 Introduction

In this chapter we describe some techniques for obtaining analytical solutions to diffusion equations in fixed domains, where the spatial boundaries are known in advance. Free boundary problems, in which the spatial boundaries vary with time in an unknown manner, are discussed in Chapter 7. We highlight in particular one method: we discuss similarity solutions in some detail. This method can yield important information about particular problems with special initial and boundary values, and it is especially useful for determining local behaviour in space or in time. It is also useful in the context of free boundary problems, and in Chapter 7 we see an application to the local behaviour of the free boundary for an American call option near expiry. Beyond this, though, we can also use similarity techniques to derive the fundamental solution of the diffusion equation, and from this we can deduce the general solution for the initial-value problem on an infinite interval. This in turn leads immediately to the Black–Scholes formulæ for the values of European call and put options. Finally, we extend the method to some options with more general payoffs, and we discuss the risk-neutral valuation method.

5.2 Similarity Solutions

It may sometimes happen that the solution $u(x, \tau)$ of a partial differential equation, together with its initial and boundary conditions, depends only on one special combination of the two independent variables. In such cases, the problem can be reduced to an ordinary differential equation in which this combination is the independent variable. The solution to this ordinary differential equation is called a **similarity solution** to the original partial differential equation. The mathematical reasons for the existence of this reduction are subtle and beyond the scope of this book, although the Technical Point at the end of this section, which deals with the mechanics of finding similarity solutions, does hint at them. We simply give two examples here.

Example 1. Suppose that $u(x,\tau)$ satisfies the following problem on the semi-infinite interval $x > 0$:

$$\frac{\partial u}{\partial \tau} = \frac{\partial^2 u}{\partial x^2}, \qquad x, \tau > 0, \tag{5.1}$$

with the initial condition

$$u(x, 0) = 0, \tag{5.2}$$

and a boundary condition at $x = 0$,

$$u(0, \tau) = 1; \tag{5.3}$$

we also require that

$$u \to 0 \text{ as } x \to \infty. \tag{5.4}$$

These equations model the evolution of temperature in a long bar, initially at zero temperature, after the temperature at one end is suddenly raised to 1 and held there.

Following the arguments suggested in the Technical Point below, we look for a solution in which $u(x,\tau)$ depends only on x and τ through the combination $\xi = x/\sqrt{\tau}$, so that $u(x,\tau) = U(\xi)$. Differentiation shows that

$$\frac{\partial u}{\partial \tau} = -\frac{1}{2\tau}\xi U'(\xi)$$

and

$$\frac{\partial^2 u}{\partial x^2} = \frac{1}{\tau}U''(\xi),$$

where $' = d/d\xi$. Substitution into equation (5.1) shows that all the terms involving τ on its own may be cancelled, and $U(\xi)$ satisfies the second order *ordinary* differential equation

$$U'' + \tfrac{1}{2}\xi U' = 0. \tag{5.5}$$

From the initial and boundary conditions (5.2)–(5.4),

$$U(0) = 1, \quad U(\infty) = 0. \tag{5.6}$$

(The second of these incorporates both (5.2) and (5.4), since as $\tau \to 0$ from above, $\xi \to \infty$.)

Separating the variables, we find that

$$U'(\xi) = Ce^{-\xi^2/4}$$

for some constant C. Integrating,

$$U(\xi) = C\int_0^\xi e^{-s^2/4}ds + D$$

where D is a further constant. Applying the boundary conditions (5.6), writing $\int_0^\xi = \int_0^\infty - \int_\xi^\infty$, and using the standard result

$$\int_0^\infty e^{-s^2/4}ds = \sqrt{\pi},$$

we find that

$$U(\xi) = \frac{1}{\sqrt{\pi}}\int_\xi^\infty e^{-s^2/4}ds;$$

that is,

$$u(x,\tau) = \frac{1}{\sqrt{\pi}}\int_{x/\sqrt{\tau}}^\infty e^{-s^2/4}ds.$$

It is easy to verify that this function does satisfy the problem statement (5.1)–(5.4), so that the solution does indeed depend only on $x/\sqrt{\tau}$.

Example 2. For our second example we derive the fundamental solution $u_\delta(x,\tau)$, which we introduced in Chapter 4. We again look for a solution of the diffusion equation that depends on x only through the combination $\xi = x/\sqrt{\tau}$, but now we try the form

$$u_\delta(x,\tau) = \tau^{-1/2}\,U_\delta(\xi).$$

The $\tau^{-1/2}$ term multiplying $U_\delta(\xi)$ is there to ensure that $\int_{-\infty}^\infty u(x,\tau)\,dx$ is constant for all τ, which can be shown by direct calculation. A computation similar to the example above shows that $U_\delta(\xi)$ satisfies the ordinary differential equation

$$U_\delta'' + (\tfrac{1}{2}\xi U_\delta)' = 0.$$

The general solution of this, obtained by integrating twice, the second time with the help of the integrating factor $e^{\xi^2/4}$, is

$$U_\delta(\xi) = Ce^{-\xi^2/4} + D$$

for constant C and D. Choosing $D = 0$ and normalising the solution by setting $C = 1/(2\sqrt{\pi})$, so that $\int_{-\infty}^{\infty} u\,dx = 1$, yields the fundamental solution

$$u_\delta(x, \tau) = \frac{1}{2\sqrt{\pi\tau}} e^{-x^2/4\tau}$$

as required.

The similarity solution technique is rarely successful in solving a complete boundary value problem, because it requires such special symmetries in the equation and the initial and boundary conditions. On the other hand, it comes into its own in local analyses in space or in time, for example the initial motion of a free boundary in an American option problem and the value of an at-the-money option shortly before exercise, which are hard to resolve numerically.

Technical Point: Group Invariances and Similarity Solutions. The key to the similarity solutions above is that both the equations and the initial and boundary conditions are invariant under the scalings $x \mapsto \lambda x$, $\tau \mapsto \lambda^2 \tau$ for any real number λ. Such a scaling is called a **one-parameter group** of transformations. This invariance is readily verified using the new variables $X = \lambda x$, $T = \lambda^2 \tau$, whereupon u is easily seen to satisfy $\partial u/\partial T = \partial^2 u/\partial X^2$. Furthermore, in Example 1, the initial and boundary conditions become $u(X, 0) = 0$, $u(0, T) = 1$ for any λ. Now $x/\sqrt{\tau} = X/\sqrt{T}$ is the only combination of X and T which is independent of λ, and so the solution must be a function of $x/\sqrt{\tau}$ only. It is essential that the equation, the boundary conditions and the initial conditions should *all* be invariant under the scaling transformation for the method to work. In Example 2, the function of τ, in this case $\tau^{-1/2}$, multiplying $U_\delta(\xi)$ is present because the diffusion equation, being linear, is also invariant under the one-parameter group $u \mapsto \mu u$. A good practical test for similarity solutions is to try $u = \tau^\alpha f(x/\tau^\beta)$ in the hope that x and τ will remain in the equations only in the combination $\xi = x/\tau^\beta$. In Example 1 above, the result of doing this is $\alpha = 0$ from the boundary condition at $x = 0$ and $\beta = \frac{1}{2}$ from the diffusion equation, while in Example 2, $\alpha = -\frac{1}{2}$ because we want the integral of $u(x, \tau)$ over x to be independent of τ, and again $\beta = \frac{1}{2}$.

5.3 An initial value problem for the diffusion equation

The fundamental solution of the diffusion equation can be used to derive an explicit solution to the initial value problem (4.3)–(4.7), in which we have to solve the diffusion equation for $-\infty < x < \infty$ and $\tau > 0$, with arbitrary initial data $u(x, 0) = u_0(x)$ and suitable growth conditions at $x = \pm\infty$. The key to the solution is the fact that we can write the initial data as

$$u_0(x) = \int_{-\infty}^{\infty} u_0(\xi)\delta(\xi - x)\, d\xi$$

where $\delta(\cdot)$ is the Dirac delta function. We recall that the fundamental solution of the diffusion equation,

$$u_\delta(s, \tau) = \frac{1}{2\sqrt{\pi t}} e^{-s^2/4\tau},$$

has initial value

$$u_\delta(s, 0) = \delta(s).$$

Now note that because $u_\delta(s - x, \tau) = u_\delta(x - s, \tau)$,

$$u_\delta(s - x, \tau) = \frac{1}{2\sqrt{\pi\tau}} e^{-(s-x)^2/4\tau}$$

is a solution of the diffusion equation using either s or x as the spatial independent variable, and its initial value is

$$u_\delta(s - x, 0) = \delta(s - x).$$

Thus, for each s, the function

$$u_0(s)u_\delta(s - x, \tau),$$

regarded as a function of x and τ with s held fixed, satisfies the diffusion equation $\partial u/\partial\tau = \partial^2 u/\partial x^2$, and has initial data $u_0(s)\delta(s - x)$. Because the diffusion equation is linear, we can superpose solutions of this form. Doing so for all s by integrating from $s = -\infty$ to $s = \infty$, we obtain a further solution of the diffusion equation,

$$u(x, \tau) = \frac{1}{2\sqrt{\pi\tau}} \int_{-\infty}^{\infty} u_0(s) e^{-(x-s)^2/4\tau} ds, \tag{5.7}$$

which has initial data

$$u(x, 0) = \int_{-\infty}^{\infty} u_0(s)\delta(s - x)\, ds = u_0(x).$$

This, therefore, is the explicit solution of the initial value problem (4.3)–(4.7). It can be shown (Exercise 1 of Chapter 4) that this solution is unique. The derivation above is not the only way of finding it: the Fourier transform is an alternative, but we do not describe it here (see any of the books referred to in Chapter 4 for treatments).

The solution (5.7) can be interpreted physically as follows. Recall that the fundamental solution of the diffusion equation describes the spreading out of a unit 'packet' of heat which, at $\tau = 0$, is all concentrated at the origin. Mathematically, this 'packet' is represented by a delta function. Now imagine the initial temperature distribution $u_0(x)$ as being made up of many small packets, the packet at $x = s$ having magnitude $u_0(s)\,ds$. Each of these evolves to give a temperature distribution equal to the fundamental solution, multiplied by $u_0(s)$ and with x replaced by $x - s$. Because the diffusion equation is linear, we obtain the whole temperature distribution by superposing (adding) the evolutions of these individual packets; in the limit, this sum is replaced by the integral (5.7).

5.4 The Black–Scholes Formulæ Derived

The Black–Scholes equation and boundary conditions for a European call with value $C(S,t)$ are, as described in Sections 3.5 and 3.6,

$$\frac{\partial C}{\partial t} + \tfrac{1}{2}\sigma^2 S^2 \frac{\partial^2 C}{\partial S^2} + rS\frac{\partial C}{\partial S} - rC = 0, \tag{5.8}$$

with

$$C(0,t) = 0, \qquad C(S,t) \sim S \quad \text{as } S \to \infty,$$

and

$$C(S,T) = \max(S - E, 0).$$

Equation (5.8) looks a little like the diffusion equation, but it has more terms, and each time C is differentiated with respect to S it is multiplied by S, giving nonconstant coefficients. Also the equation is clearly in backward form, with final data given at $t = T$.

The first thing to do is to get rid of the awkward S and S^2 terms multiplying $\partial C/\partial S$ and $\partial^2 C/\partial S^2$. At the same time we take the opportunity of making the equation **dimensionless**, as defined in the Technical Point below, and we turn it into a forward equation. We set

$$S = Ee^x, \quad t = T - \tau/\tfrac{1}{2}\sigma^2, \quad C = Ev(x,\tau). \tag{5.9}$$

This results in the equation

$$\frac{\partial v}{\partial \tau} = \frac{\partial^2 v}{\partial x^2} + (k-1)\frac{\partial v}{\partial x} - kv \tag{5.10}$$

where $k = r/\frac{1}{2}\sigma^2$. The initial condition becomes

$$v(x,0) = \max(e^x - 1, 0).$$

Notice in particular that this equation contains only *one* dimensionless parameter, $k = r/\frac{1}{2}\sigma^2$, although there are *four* dimensional parameters, E, T, σ^2 and r, in the original statement of the problem. There is in fact another, $\frac{1}{2}\sigma^2 T$, the dimensionless time to expiry, and these two are the only genuinely independent parameters in the problem; the effect of all other factors is simply brought in by inverting the above transformations, i.e. by a straightforward arithmetical calculation.

Equation (5.10) now looks much more like a diffusion equation, and we can turn it into one by a simple change of variable. If we try putting

$$v = e^{\alpha x + \beta \tau} u(x,\tau),$$

for some constants α and β to be found, then differentiation gives

$$\beta u + \frac{\partial u}{\partial \tau} = \alpha^2 u + 2\alpha \frac{\partial u}{\partial x} + \frac{\partial^2 u}{\partial x^2} + (k-1)\left(\alpha u + \frac{\partial u}{\partial x}\right) - ku.$$

We can obtain an equation with no u term by choosing

$$\beta = \alpha^2 + (k-1)\alpha - k,$$

while the choice

$$0 = 2\alpha + (k-1)$$

eliminates the $\partial u/\partial x$ term as well. These equations for α and β give

$$\alpha = -\tfrac{1}{2}(k-1), \quad \beta = -\tfrac{1}{4}(k+1)^2.$$

We then have

$$v = e^{-\frac{1}{2}(k-1)x - \frac{1}{4}(k+1)^2\tau} u(x,\tau),$$

where

$$\frac{\partial u}{\partial \tau} = \frac{\partial^2 u}{\partial x^2} \quad \text{for } -\infty < x < \infty,\ \tau > 0,$$

with

$$u(x,0) = u_0(x) = \max\left(e^{\frac{1}{2}(k+1)x} - e^{\frac{1}{2}(k-1)x}, 0\right). \tag{5.11}$$

This may seem like a long way to travel from the original formulation, but we have reached the payoff. The solution to the diffusion equation problem is just that given in equation (5.7):

$$u(x,\tau) = \frac{1}{2\sqrt{\pi\tau}} \int_{-\infty}^{\infty} u_0(s) e^{-(x-s)^2/4\tau} \, ds \tag{5.12}$$

where $u_0(x)$ is given by (5.11).

It remains to evaluate the integral in (5.12). It is convenient to make the change of variable $x' = (s-x)/\sqrt{2\tau}$, so that

$$\begin{aligned} u(x,\tau) &= \frac{1}{\sqrt{2\pi}} \int_{-\infty}^{\infty} u_0(x'\sqrt{2\tau}+x) e^{-\frac{1}{2}x'^2} \, dx' \\ &= \frac{1}{\sqrt{2\pi}} \int_{-x/\sqrt{2\tau}}^{\infty} e^{\frac{1}{2}(k+1)(x+x'\sqrt{2\tau})} e^{-\frac{1}{2}x'^2} \, dx' \\ &\quad - \frac{1}{\sqrt{2\pi}} \int_{-x/\sqrt{2\tau}}^{\infty} e^{\frac{1}{2}(k-1)(x+x'\sqrt{2\tau})} e^{-\frac{1}{2}x'^2} \, dx' \\ &= I_1 - I_2, \end{aligned}$$

say.

We evaluate I_1 by completing the square in the exponent to get a standard integral:

$$\begin{aligned} I_1 &= \frac{1}{\sqrt{2\pi}} \int_{-x/\sqrt{2\tau}}^{\infty} e^{\frac{1}{2}(k+1)(x+x'\sqrt{2\tau})-\frac{1}{2}x'^2} \, dx' \\ &= \frac{e^{\frac{1}{2}(k+1)x}}{\sqrt{2\pi}} \int_{-x/\sqrt{2\tau}}^{\infty} e^{\frac{1}{4}(k+1)^2\tau} e^{-\frac{1}{2}\left(x'-\frac{1}{2}(k+1)\sqrt{2\tau}\right)^2} \, dx' \\ &= \frac{e^{\frac{1}{2}(k+1)x+\frac{1}{4}(k+1)^2\tau}}{\sqrt{2\pi}} \int_{-x/\sqrt{2\tau}-\frac{1}{2}(k+1)\sqrt{2\tau}}^{\infty} e^{-\frac{1}{2}\rho^2} \, d\rho \\ &= e^{\frac{1}{2}(k+1)x+\frac{1}{4}(k+1)^2\tau} N(d_1), \end{aligned}$$

where

$$d_1 = \frac{x}{\sqrt{2\tau}} + \tfrac{1}{2}(k+1)\sqrt{2\tau},$$

and

$$N(d_1) = \frac{1}{\sqrt{2\pi}} \int_{-\infty}^{d_1} e^{-\frac{1}{2}s^2} \, ds$$

is the cumulative distribution function for the normal distribution.

The calculation of I_2 is identical to that of I_1, except that $(k+1)$ is replaced by $(k-1)$ throughout.

Lastly, we retrace our steps, writing

$$v(x,\tau) = e^{-\frac{1}{2}(k-1)x-\frac{1}{4}(k+1)^2\tau}u(x,\tau) \tag{5.13}$$

and then putting $x = \log(S/E)$, $\tau = \frac{1}{2}\sigma^2(T-t)$ and $C = Ev(x,\tau)$, to recover

$$C(S,t) = SN(d_1) - Ee^{-r(T-t)}N(d_2), \tag{5.14}$$

where

$$\begin{aligned} d_1 &= \frac{\log(S/E) + (r+\frac{1}{2}\sigma^2)(T-t)}{\sigma\sqrt{(T-t)}}, \\ d_2 &= \frac{\log(S/E) + (r-\frac{1}{2}\sigma^2)(T-t)}{\sigma\sqrt{(T-t)}}. \end{aligned}$$

The corresponding calculation for a European put option follows similar lines. Its transformed payoff is

$$u(x,0) = \max\left(e^{\frac{1}{2}(k-1)x} - e^{\frac{1}{2}(k+1)x}, 0\right), \tag{5.15}$$

and we can proceed as above. However, having evaluated the call, a simpler way is to use the put-call parity formula

$$C - P = S - Ee^{-r(T-t)}$$

for the value P of a put given the value of the call. This yields

$$P(S,t) = Ee^{-r(T-t)}N(-d_2) - SN(-d_1),$$

where we have used the identity $N(d) + N(-d) = 1$.

The deltas of call and put options are calculated by differentiation: for the call,

$$\begin{aligned} \Delta &= \frac{\partial C}{\partial S} \\ &= N(d_1) + S\frac{\partial}{\partial S}N(d_1) - Ee^{-r(T-t)}\frac{\partial}{\partial S}N(d_2) \\ &= N(d_1) + SN'(d_1)\frac{\partial d_1}{\partial S} - Ee^{-r(T-t)}N'(d_2)\frac{\partial d_2}{\partial S} \\ &= N(d_1) + \left(SN'(d_1) - Ee^{-r(T-t)}N'(d_2)\right) / \, S\sigma\sqrt{T-t} \\ &= N(d_1), \end{aligned}$$

since a rather painful calculation shows that $SN'(d_1) = Ee^{-r(T-t)}N'(d_2)$ (divide both sides by $N'(d_2) = (1/\sqrt{2\pi})e^{-\frac{1}{2}d_2^2}$ first). Then, the delta for the put is

$$\begin{aligned} \Delta &= \frac{\partial P}{\partial S} \\ &= N(d_1) - 1, \end{aligned}$$

again using put-call parity. These quantities are vital if an option position is to be hedged correctly.

Some computer algebra packages offer a limited range of financial routines. Maple, for example, has a Black–Scholes call command. It has to be loaded by typing

```
>readlib(finance);
```

and then the command

```
>blackscholes(E,T-t,S,sigma,r);
```

returns the Black–Scholes value of the call with exercise price `E`, time to expiry `T-t`, current asset price `S`, interest rate `r` and volatility `sigma`. In this example the symbols have to be replaced by their numerical values, but the routine can also be used as a function. For example,

```
>plot(blackscholes(10,0.5,S,0.2,0.1), S=0..20);
```

generates a plot of the call values for $0 < S < 20$, with the other parameters held fixed at the values indicated. Other Maple features can also be used; for example

```
>plot(diff(blackscholes(10,0.5,S,0.2,0.1),S),S=0..20);
>plot(diff(blackscholes(10,0.5,S,0.2,0.1),S$2),S=0..20);
```

plot the delta and gamma respectively. Although there is no separate put routine, it is easy to write one using the call routine and put-call parity.

Technical Point: Dimensionless Variables.

The differential equations used to model physical and financial processes often contain many **parameters**; these might be material properties of the substances involved, for example, thermal conductivity, or constants of the underlying stochastic processes, such as their rate of return or volatility. An early step in most solutions is to scale the dependent and independent variables with 'typical values' in order to collect these parameters together as far as possible. Thus above we scaled S and V with E, the only *a priori* typical value available. Although S might be measured in £ (or \$, or DM, or any other units), x has no units, and nor does v. This is important, since an expansion of the form $e^S = 1+S+\frac{1}{2}S^2+\cdots$ is meaningless if S is a dimensional quantity. (Note that an absolute change in an asset value, dS, is dimensional, but that the relative change, dS/S, is not.)

Having carried out this scaling, we can collect the remaining parameters into **dimensionless groups**, also called **dimensionless parameters**. This scaling tells us the true number of independent constants in the solution. If one of the resulting dimensionless parameters is very large or very small, we may subsequently be able to exploit this fact to construct a useful approximation to the solution. Such an approximation is called an **asymptotic expansion**, and the theory of **asymptotic analysis** aims to devise techniques for this kind of approximation. It also aims to analyse the techniques in order to make sure that we can be confident that the effects we have neglected in making the approximation are genuinely unimportant.

In the Black–Scholes equation, both r and σ^2 have units $(\text{time})^{-1}$; $(\text{years})^{-1}$ or $(\text{days})^{-1}$, for example. The quantity $k = r/\frac{1}{2}\sigma^2$ is a dimensionless parameter. Another is $\frac{1}{2}\sigma^2 T$, the dimensionless lifetime. These are the only dimensionless parameters in the basic problem for a European call or put.

5.5 Binary Options

Although we discussed only vanilla calls and puts in the previous section, it was only at the very last stage that we needed to know which option we were dealing with. The function $u_0(s)$ in equation (5.7) can clearly be the payoff for *any* combination of options: the linearity of the Black–Scholes equation guarantees that we can value portfolios of options by superposition. In this way, we can value combinations such as straddles, strangles and so on. Furthermore, the payoff need not be a finite combination of calls and puts: we can consider any function of S that we wish. Options with payoffs more general than vanilla calls and puts are known as **binary options** or **digital options**.

Suppose that the payoff at time T is $\Lambda(S)$, and that the value of the option is $V(S,t)$, so $V(S,T) = \Lambda(S)$. We first work out the function $u_0(x)$ corresponding to $\Lambda(S)$ after the transformations that we used above. That is, we set $S = Ee^x$ and then $V(S,t) = Ee^{\alpha x + \beta \tau} u(x,\tau)$, where α, β and τ have their previous meanings. From the payoff, $V(S,T) = \Lambda(S) = Ee^{\alpha x} u_0(x)$. Then, from equation (5.7) we have a formula for $u(x,\tau)$; undoing the changes of variable leads to the explicit formula

$$\frac{e^{-r(T-t)}}{\sigma\sqrt{2\pi(T-t)}} \int_0^\infty \Lambda(S') e^{-\left(\log(S'/S)-(r-\frac{1}{2}\sigma^2)(T-t)\right)^2/2\sigma^2(T-t)} \frac{dS'}{S'}, \tag{5.16}$$

for $V(S,t)$. This formula obviously includes vanilla calls and puts as particular cases. The delta is given by the derivative of (5.16) with respect to S. In deriving (5.16), we have assumed that σ and r are constant and that the underlying pays no dividends. The inclusion of a dividend term is not difficult, and if σ or r is a known function of t then the methods described in Chapter 6 may be applied to obtain exact formulæ.

One particularly popular binary option has already been mentioned: the cash-or-nothing call, whose payoff is

$$\Lambda(S) = B\mathcal{H}(S-E).$$

This option can be interpreted as a simple bet on the asset price; if $S > E$ at expiry the payoff is B and otherwise it is zero. (More often, though, it is found as part of a 'structured product' with conditions that allow for a fixed payment to be made if an asset is above a certain value on a certain date.) Its value is

$$V(S,t) = Be^{-r(T-t)} N(d_2),$$

where d_2 is as above. Another binary option, sometimes known as a **supershare**, has payoff $1/d$ if $E < S < E + d$ at expiry and zero otherwise:

$$\Lambda(S) = \frac{1}{d}\Big(\mathcal{H}(S-E) - \mathcal{H}(S-E-d)\Big)$$

(in the limit $d \to 0$ the payoff becomes a delta function). Its valuation is left as an exercise.

Although these options are easy to value using (5.16) they can present problems in hedging near the time of expiry, caused by the discontinuities in the payoff function. Consider, for example, the difficulties associated with hedging a cash-or-nothing call with payoff $B\mathcal{H}(S-E)$. By differentiating $\mathcal{H}(S-E)$ with respect to S we see that as $t \to T$ the delta of the option tends to the function $B\delta(S-E)$. Away from $S = E$ this function is zero, and therefore close to expiry we expect that we should not have to hedge the option. However, if S is close to E near expiry there is a high probability that the asset price will cross the value E, perhaps many times, before expiry. Each time this value is crossed the delta goes from nearly zero to very large and back to nearly zero. The Black–Scholes model assumes that the option is continuously hedged with a number of assets equal to the delta; this is clearly impractical if, at one moment, the portfolio contains no assets, then is rehedged to contain a large number of the assets only for that position to be liquidated shortly afterwards. Yet, if this rehedging is not done, the payoff at expiry is either zero or B, and cannot be known for certain.[1] It is therefore open to question whether options with discontinuous payoffs can be valued according to the simple Black–Scholes formula (5.16).

5.6 Risk Neutrality

A rather different view of option valuation from that presented above is the **risk-neutral** approach. This stems from the observation that the growth rate μ does not appear in the Black–Scholes equation (3.9). Therefore, although the value of an option depends on the standard deviation of the asset price, it does not depend on its rate of growth. Indeed, different investors may have widely varying estimates of the growth rate of a share yet still agree on the value of an option. Moreover, the risk preferences of investors are irrelevant: because the risk inherent in an option can all be hedged away, there is no return to be made over and above the risk-free return. Whether for vanilla options or other products, it is generally the case that if a portfolio can be constructed with a derivative product and the underlying asset in such a way that the random component can be eliminated – as was the case in our derivation of the Black–Scholes equation in Chapter 3 – then the derivative product may be valued as if *all* the random walks involved are **risk-neutral**. This means that the drift term in the stochastic differential equation for the asset return (for our equity model, μ) is replaced by r wherever it appears. The option is then valued by calculating the present value of its *expected* return at expiry with this modification to the random walk. The process works as follows.

We begin by recalling that the present value of any amount at time T

[1] There is a similar but less important effect close to expiry for vanilla options if the asset price is close to the exercise price.

is that amount discounted by multiplying by $e^{-r(T-t)}$. Then, we set up a risk-neutral world: we pretend that the random walk for the return on S has drift r instead of μ. From this, we can calculate the probability density function of future values of S: we use equation (2.10) with μ replaced by r. *It is most important to realise that the new probability density function is* not *that of* S. Next, we calculate the expected value of the payoff $\Lambda(S)$ using this probability density function. That is, we multiply $\Lambda(S)$ by the risk-neutral probability density function and integrate over all possible future values of the asset, from zero to infinity. Finally, we discount to get the present value of the option. The resulting formula is, as before,

$$V(S,t) = \frac{e^{-r(T-t)}}{\sigma\sqrt{2\pi(T-t)}} \times \int_0^\infty e^{-\left(\log(S'/S)-(r-\frac{1}{2}\sigma^2)(T-t)\right)^2/2\sigma^2(T-t)} \Lambda(S') \frac{dS'}{S'}. \tag{5.17}$$

This expression can be shown by direct differentiation to satisfy equation (3.9). When the payoff is simple, it can be integrated explicitly to give the Black–Scholes formula for (for example) a European call option.

The idea of replacing μ by r is very elegant. It does, however, have some major drawbacks. First, it requires us to know the probability density function of the future asset values (under the risk-neutral assumption). This is easy enough for our constant-coefficient random walks, but if we want to use any more complicated model, we must first find the distribution before integrating to calculate the expected return. Often, the calculation of the probability density function involves solving a partial differential equation equivalent to that satisfied by the option, and the subsequent integration must in general be carried out numerically as well. It is usually quicker to solve the option pricing equation directly. Moreover, when we come to exotic options or American options, it is much more difficult to see how to implement the risk-neutral approach, while (as we show) the direct approach via the partial differential equation for the option can be extended in a clear-cut way.

A further drawback is that risk neutrality can lead to confusion. For example, it is sometimes said that

- "It can be shown that $\mu = r$,"

or that

- "The delta of an option is the probability that it will expire in the money."

Both of these statements are wrong. If the first statement were correct then all assets would have the same expected return as a bank deposit and no one would invest in equities (see the Technical Point on risk in Chapter 2). If μ were equal to r then the second statement would be correct. The probability that $S > E$ at $t = T$ can be found by calculating the expected value of $\mathcal{H}(S - E)$. This necessarily involves the parameter μ.

Finally, risk-neutrality is far from easy to grasp intuitively, which is perhaps the source of the confusion above. The key steps in the derivation of the Black–Scholes equation, namely no arbitrage and that risk-free portfolios earn the risk-free rate, are intuitively clear.

Further Reading

- For a discussion of similarity solutions of the diffusion equation see Crank (1989) and Hill & Dewynne (1990).
- The risk-neutral method is set out in the papers by Cox & Ross (1976) and Harrison & Pliska (1981). For some details of option valuation under risk neutrality, see Harrison & Kreps (1979) and Hull (1993).

Exercises

1. Find a similarity solution to the problem

$$\frac{\partial u}{\partial \tau} = \frac{\partial^2 u}{\partial x^2}, \quad -\infty < x < \infty, \quad \tau > 0,$$

with

$$u(x, 0) = \mathcal{H}(x).$$

Show that $\partial u/\partial x$ is the fundamental solution $u_\delta(x, \tau)$, either by direct differentiation or by constructing the initial value problem that it satisfies.

2. Suppose that $u(x, \tau)$ satisfies the following initial value problem on a semi-infinite interval:

$$\frac{\partial u}{\partial \tau} = \frac{\partial^2 u}{\partial x^2}, \quad x > 0, \ \tau > 0,$$

with

$$u(x,0) = u_0(x),\ x > 0, \quad u(0,\tau) = 0,\ \tau > 0.$$

Define a new function $v(x,\tau)$ by reflection in the line $x = 0$, so that

$$v(x,\tau) = u(x,\tau) \quad \text{if } x > 0,$$

$$v(x,\tau) = -u(-x,\tau) \quad \text{if } x < 0.$$

Show that $v(0,\tau) = 0$, and use (5.7) to show that

$$u(x,\tau) = \frac{1}{2\sqrt{\pi\tau}} \int_0^\infty u_0(s) \left(e^{-(x-s)^2/4\tau} - e^{-(x+s)^2/4\tau} \right) ds.$$

The function multiplying $u_0(s)$ here is called the **Green's function** for this initial-boundary value problem. This solution is applicable to barrier options.

3. Find similarity solutions to

$$\frac{\partial u}{\partial \tau} = \frac{\partial^2 u}{\partial x^2} + F(x), \quad x > 0,\ \tau > 0,$$

with

$$u(x,0) = 0,\ x > 0, \quad u(0,t) = 0,\ \tau > 0.$$

in the two cases (a) $F(x) = x$; (b) $F(x) = 1$.

Extend case (b) by letting $u(0,\tau) = \tau$. A related similarity solution plays an important role in the free boundary problems studied in Chapter 7.

4. Suppose that a and b are constants. Show that the parabolic equation

$$\frac{\partial u}{\partial \tau} = \frac{\partial^2 u}{\partial x^2} + a\frac{\partial u}{\partial x} + bu$$

can always be reduced to the diffusion equation. Use a change of time variable to show that the same is true for the equation

$$c(\tau)\frac{\partial u}{\partial \tau} = \frac{\partial^2 u}{\partial x^2}$$

where $c(\tau) > 0$. Suppose that σ^2 and r in the Black–Scholes equation are both functions of t, but that r/σ^2 is constant. Derive the Black–Scholes formulæ in this case.

5. Suppose that in the Black–Scholes equation, $r(t)$ and $\sigma^2(t)$ are both non-constant but known functions of t. Show that the following procedure reduces the Black–Scholes equation to the diffusion equation.

(a) Set $S = Ee^x$, $C = Ev$ as before, and put $t = T - t'$ to get the equation

$$\frac{\partial v}{\partial t'} = \tfrac{1}{2}\sigma^2(t')\frac{\partial^2 v}{\partial x^2} + \left(r(t') - \tfrac{1}{2}\sigma^2(t')\right)\frac{\partial v}{\partial x} - r(t')v.$$

Note that we have not yet scaled time, but merely changed its origin.

(b) Now introduce a new time variable $\hat{\tau}$ such that $\frac{1}{2}\sigma^2(t')dt' = d\hat{\tau}$, i.e.

$$\hat{\tau}(t') = \int_0^{t'} \tfrac{1}{2}\sigma^2(s)\, ds.$$

(See the previous question where this calculation is requested.) This change of time variable amounts to measuring time weighted by volatility, so that the new 'time' passes more slowly when the volatility is high. The resulting equation is

$$\frac{\partial v}{\partial \hat{\tau}} = \frac{\partial^2 v}{\partial x^2} + a(\hat{\tau})\frac{\partial v}{\partial x} - b(\hat{\tau})v,$$

where $a(\hat{\tau}) = r/\frac{1}{2}\sigma^2 - 1$, $b(\hat{\tau}) = r/\frac{1}{2}\sigma^2$ (note that the dependence of r and σ^2 on $\hat{\tau}$ is obtained by substituting for t' in terms of $\hat{\tau}$ by inverting the change of variable above).

(c) Show (or verify) that the general solution of the *first* order partial differential equation obtained from the equation for v by omitting the term $\partial^2 v/\partial x^2$, namely

$$\frac{\partial v}{\partial \hat{\tau}} = a(\hat{\tau})\frac{\partial v}{\partial x} - b(\hat{\tau})v,$$

is

$$v(x, \hat{\tau}) = F\big(x + A(\hat{\tau})\big)e^{-B(\hat{\tau})},$$

where $dA/d\hat{\tau} = a(\hat{\tau})$ and $dB/d\hat{\tau} = b(\hat{\tau})$, and $F(\cdot)$ is an arbitrary function.

(d) Now seek a solution to the full equation for v in the form

$$v(\hat{x}, \hat{\tau}) = e^{-B(\hat{\tau})}V(x, \hat{\tau}),$$

where $\hat{x} = x + A(\hat{\tau})$ is as above. Choose $B(\hat{\tau})$ so that V satisfies the diffusion equation

$$\frac{\partial V}{\partial \hat{\tau}} = \frac{\partial^2 V}{\partial x^2}.$$

(e) What happens to the initial data under this series of transformations?

See Harper (1993) for further examples of this ingenious procedure applied to other equations.

6. Show that equation (5.10) can also be reduced to the diffusion equation by writing

$$v(x,\tau) = e^{-k\tau} V(\xi,\tau),$$

where

$$\xi = x + (k-1)\tau.$$

What disadvantages might there be to this change of variables?

7. If $C(S,t)$ and $P(S,t)$ are the values of a European call and put with the same exercise and expiry, show that $C - P$ also satisfies the Black–Scholes equation (5.8), with the particularly simple final data $C - P = S - E$ at $t = T$. Deduce from the put-call parity theorem that $S - Ee^{-r(T-t)}$ is also a solution; interpret these results financially.

8. Use the explicit solution of the diffusion equation to derive the Black–Scholes value for a European put option without using put-call parity.

9. Calculate the gamma, theta, vega and rho for European call and put options.

10. Use Maple (or any other computer algebra package) to plot out the functions of Exercise 9. Use the `plot3d` command to generate three-dimensional plots of call and put options as functions of two variables, for example, S and t or S and σ.

11. What is the random walk followed by a European call option?

12. If $u(x,0)$ in the initial value problem for the heat equation on an infinite interval (equations (4.3)–(4.7)) is positive, then so is $u(x,\tau)$ for $\tau > 0$. Show this, and deduce that any option whose payoff is positive always has a positive value.

13. Consider the following initial value problem on an infinite interval:

$$\frac{\partial u}{\partial \tau} = \frac{\partial^2 u}{\partial x^2} + f(x,\tau)$$

with

$$u(x,0) = 0 \quad \text{and} \quad u \to 0 \quad \text{as} \quad x \to \pm\infty.$$

It can be shown (for example by the Green's function representation of the solution) that if $f(x,\tau) \geq 0$ then $u(x,\tau) \geq 0$. Why is this physically reasonable? Use this result to show that if $C_1(S,t)$ and $C_2(S,t)$ are the values of two otherwise identical calls with different volatilities σ_1 and $\sigma_2 < \sigma_1$, then $C_2 < C_1$. Is the same result true for puts?

14. In dimensional variables, heat conduction in a bar of length L is modelled by

$$\rho c \frac{\partial U}{\partial T} = k \frac{\partial^2 U}{\partial X^2}$$

for $0 < X < L$, where $U(X, T)$ is the dimensional temperature, ρ is the density, c is the specific heat, and k is the thermal conductivity. Suppose also that U_0 is a typical value for temperature variations, either of the initial temperature $U_0(X)$, or of the boundary values at $X = 0, L$; make the equation dimensionless.

15. What is the value of an option with payoff $\mathcal{H}(E - S)$? What is the value of a supershare?

16. The European **asset-or-nothing** call pays S if $S > E$ at expiry, and nothing if $S \leq E$. What is its value?

17. What *is* the probability that a European call will expire in-the-money?

18. An option has a general payoff $\Lambda(S)$ at time T, and its value is $V(S, t)$. Show how to synthesise it from vanilla call options with varying exercise prices; that is, how to find the 'density' $f(E)$ of calls, with the same expiry T, exercise price E and price $C(S, t; E)$, such that

$$V(S, t) = \int_0^\infty f(E) C(S, t; E)\, dE.$$

Verify that your answer is correct

(a) when $\Lambda(S) = \max(S - E, 0)$ (a vanilla call);
(b) when $\Lambda(S) = S$. (What is the synthesizing portfolio here?)

Repeat the exercise using cash-or-nothing calls as the basis.

19. Suppose that European calls of all exercise prices are available. Regarding S as fixed and E as variable, show that their price $C(E, t)$ satisfies the partial differential equation

$$\frac{\partial C}{\partial t} + \tfrac{1}{2}\sigma^2 E^2 \frac{\partial^2 C}{\partial E^2} - rE \frac{\partial C}{\partial E} = 0.$$

20. "If an asset has zero volatility, then its future path is deterministic, and specified completely by μ. Therefore, we can calculate exactly the value of a call option on the asset, and it too must depend explicitly on μ. However, it is repeatedly stated above that this is *not* the case." Why is this not a contradiction?

6 Variations on the Black–Scholes Model

6.1 Introduction

We have now completed the Black–Scholes analysis of vanilla European call and put options. Although the formulæ that we have derived are useful, there are many more complicated situations in which they are not adequate. This chapter is devoted to a number of straightforward extensions of the Black–Scholes analysis. We see how to incorporate dividends, how to deal with forward and futures contracts, and how to put time-varying parameters into the Black–Scholes equation, but we still use straightforward calls and puts as the building blocks. Later chapters deal with American and 'exotic' options which have more complex contract structures.

There is one possible direction of generalisation that we do not discuss in this book: we assume that all our models are driven by stochastic processes of the type discussed previously. We do not use models that, for example, postulate some essential nonlinearity in the underlying markets, as might be attributed to feedback from derivatives markets into asset prices. Although there is some evidence that markets are not as close to our models as we would like, the Black–Scholes world is a good enough approximation for most purposes, both theoretical and practical.

6.2 Options on Dividend-paying Assets

6.2.1 Dividend Structures

Many assets, such as equities, pay out **dividends**. These are payments to shareholders out of the profits made by the company concerned, and the likely future dividend stream of a company is reflected in today's share price. The price of an option on an underlying asset that pays dividends is affected by the payments, so we must modify the Black–Scholes analysis.

When we model dividend payments, we need to consider two issues:

- When, and how often, are dividend payments made?
- How large are the payments?

There are several possible different structures for dividend payments. Individual companies usually make two or four payments per year, which may need to be treated discretely, but the large number of dividend payments on an index such as the *S&P 500* are so frequent that it may be best to regard them as a continuous payment rather than as a succession of discrete payments. Another example where dividends can be modelled as continuous is when the asset is a foreign currency, in which case the 'dividend' represents payments at the foreign interest rate (we assume for now that this is constant).

The amounts paid as dividends may be modelled as either deterministic or stochastic. In this book we consider only deterministic dividends, whose amount and timing are known at the start of an option's life. This is a reasonable assumption, since many companies endeavour to maintain a similar dividend policy from year to year.

6.2.2 A Constant Dividend Yield

Let us consider the very simplest payment structure. Suppose that in a time dt the underlying asset pays out a dividend $D_0 S\,dt$ where D_0 is a constant. This payment is independent of time except through the dependence on S. The **dividend yield** is defined as the proportion of the asset price paid out per unit time in this way. Thus the dividend $D_0 S\,dt$ represents a constant and continuous dividend yield D_0. This dividend structure is a good model for index options and for short-dated currency options (it is debatable whether (2.1) is a good model for currencies over long timescales). In the latter case $D_0 = r_f$, the foreign interest rate.

First, we consider the effect of the dividend payments on the asset price. Arbitrage considerations show that in each time-step dt, the asset price must fall by the amount of the dividend payment, $D_0\,dt$, in addition to the usual fluctuations. It follows that the random walk for the asset price (2.1) is modified to

$$dS = \sigma S\,dX + (\mu - D_0)S\,dt. \tag{6.1}$$

We have seen that the Black–Scholes equation is unaffected by the coefficient of dt in the stochastic differential equation for S and so one might expect the dividend to have no effect on the option price. This is not the case. We have allowed for the effect of the dividend payment on the asset price but not its effect on the value of our hedged portfolio. Since we receive $D_0 S\,dt$ for every asset held and since we hold $-\Delta$ of the underlying, our portfolio changes by an amount

$$-D_0 S\Delta\,dt, \tag{6.2}$$

i.e. the dividend our assets receive. Thus, we must add (6.2) to our earlier $d\Pi$ to arrive at

$$d\Pi = dV - \Delta\,dS - D_0 S\Delta\,dt.$$

The analysis proceeds exactly as before but with the addition of this new term. We find that

$$\frac{\partial V}{\partial t} + \tfrac{1}{2}\sigma^2 S^2 \frac{\partial^2 V}{\partial S^2} + (r - D_0)S\frac{\partial V}{\partial S} - rV = 0. \tag{6.3}$$

For a call option the final condition is still $C(S,T) = \max(S-E,0)$, and the boundary condition at $S=0$ remains as $C(0,t)=0$. The only change to the boundary conditions when we use the modified Black–Scholes equation (6.3) is that

$$C(S,t) \sim Se^{-D_0(T-t)} \quad \text{as} \quad S \to \infty. \tag{6.4}$$

This is because in the limit $S \to \infty$, the option becomes equivalent to the asset *but without its dividend income.*

We could calculate the value of this option in the same way as we did without dividends: that is, reduce equation (6.3) to the diffusion equation and solve in the usual way. However, it is quicker to notice that we can make the coefficients of $S\,\partial C/\partial S$ and C in (6.3) equal by setting

$$C(S,t) = e^{-D_0(T-t)}C_1(S,t).$$

We then see that $C_1(S,t)$ satisfies the basic Black–Scholes equation (3.9) *with r replaced by $r - D_0$ and with the same final value.* The value of $C_1(S,t)$ is therefore just that of a normal European call with interest rate $r - D_0$, and it is now straightforward to show that with dividends, the value of a European call option is

$$C(S,t) = e^{-D_0(T-t)}SN(d_{10}) - Ee^{-r(T-t)}N(d_{20}),$$

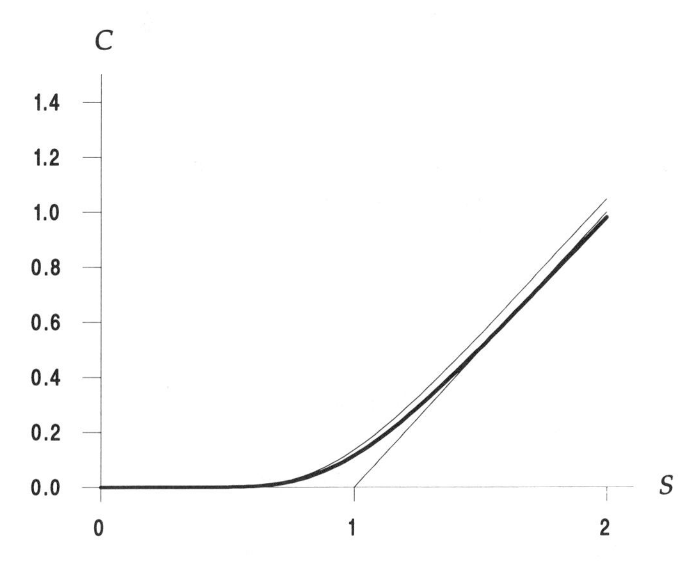

Figure 6.1 A comparison of European call option values with (lower curve) and without dividends (upper curve). There are six months to expiry, $E = 1$, $\sigma = 0.4$ and $r = 0.1$. The bold curve has $D_0 = 0.07$.

where

$$d_{10} = \frac{\log(S/E) + (r - D_0 + \frac{1}{2}\sigma^2)(T-t)}{\sigma\sqrt{T-t}}, \quad d_{20} = d_1 - \sigma\sqrt{T-t}.$$

(Alternatively, we can use the variable $\hat{S} = Se^{D_0 t}$; the value of the call is the usual Black–Scholes value with S replaced by $\hat{S}e^{-D_0(T-t)}$ throughout.)

In Figure 6.1 we see the European call option values as functions of S with six months to expiry, $\sigma = 0.4$ and $r = 0.1$; the top curve is the value of the option in the absence of dividends, and the lower bold curve has a constant and continuous dividend yield $D_0 = 0.07$.

Further properties of options on assets with a continuous dividend yield are developed in the exercises at the end of the chapter.

6.2.3 Discrete Dividend Payments

Suppose that our asset pays just one dividend during the lifetime of the option, at time $t = t_d$. As above, we shall consider only the case in which the dividend *yield* is a known constant d_y (obviously, $0 \leq d_y \leq 1$; usually it is a few percent at most). Thus, at time t_d, holders of the asset receive a payment $d_y S$, where S is the asset price just before the dividend is paid.

First, consider the effect of the dividend payment on the asset price. Its value just before[1] the dividend date, at time t_d^-, cannot equal its value just after, at time t_d^+. If it did, the strategy of buying the asset immediately before t_d, collecting the dividend, and selling straight away, would yield a risk-free profit. It is clear that, in the absence of other factors such as taxes, the asset price must fall by exactly the amount of the dividend payment. Thus,

$$S(t_d^+) = S(t_d^-) - d_y S(t_d^-) = S(t_d^-)(1 - d_y). \tag{6.5}$$

We now have to incorporate this jump into our model for options.

The ideas we present below are important not only for discrete dividends, but also for many exotic options with discrete contract features. They should be read with care.

6.2.4 Jump Conditions for Discrete Dividends

We have just seen that a discrete dividend payment inevitably results in a jump in the value of the underlying asset across the dividend date. Our next task is to determine what effect the jump has on the option price. This brings us to the subject of **jump conditions**.

Jump conditions arise when there is a discontinuous change in one of the independent variables affecting the value of a derivative security. Here, the cause of a jump is the discontinuous change in asset price due to the discrete payment of a dividend, but later we shall deal with other causes in connection with exotic options. The jump condition relates the values of the option across the jump; in this case it relates the values of the option before and after the dividend date.

Jump conditions may be derived in two equivalent ways. One method is via financial arguments, and is based on arbitrage considerations. The other way is a purely mathematical method, based on the manipulation of delta functions and first order hyperbolic partial differential equations. We present the financial argument here; for the mathematical arguments, see the Technical Point at the end of this section and *Option Pricing*.

Away from the dividend date the value of the option varies because of the random movement of the asset price; this variation is gradual in

[1] We shall consistently use the notation t^- and t^+ to denote the moments just before and just after time t.

time, since the movement of the asset price is continuous in time (albeit random). Across the dividend date, however, the value of the asset changes discontinuously. This change in asset price is given by (6.5).

Now consider the effect of this discontinuous change in the asset value, S, on an option, with value $V(S,t)$, contingent on that asset. To eliminate the same sort of arbitrage possibilities as those considered above, *the value of the option must be continuous as a function of time across the dividend date*; the value of the option is the same immediately before the dividend date as it is immediately after (recall that the holder of the option does not receive the dividend). Thus we arrive at the jump condition

$$V\left(S(t_d^-), t_d^-\right) = V\left(S(t_d^+), t_d^+\right). \tag{6.6}$$

This jump condition arises from eliminating arbitrage possibilities *for any given realisation of the asset and option values.* That is, the option value must be continuous in time for any realisation of the asset's random walk. We have just asserted that the option price is continuous in time, yet we have called this a jump condition, which implies discontinuity. How can we reconcile these two statements?

In this book, we analyse option models using partial differential equations with S and t as independent variables. We do this instead of thinking of S as a function of t, as is implicit in (6.6), because we need to be able to consider all possible realisations of the asset's random walk. Bearing this in mind, let us now consider what happens to the option value across a dividend date in a Black–Scholes model. Since we regard S and t as independent variables in such a formulation, it would seem at first sight that this question could be phrased as

- How does V change across a dividend date for S fixed?

In fact, in any realisation, S would not be fixed across a dividend date. The question we have just posed is not quite appropriate to the problem, and it is better to ask

- How does V change as a function of S across a dividend date?

The answer is that V changes discontinuously according to (6.6) with $S(t_d^+)$ and $S(t_d^-)$ related by (6.5). That is, we have

$$V\left(S, t_d^-\right) = V\left(S(1-d_y), t_d^+\right). \tag{6.7}$$

This says that the value of the option at asset value S immediately before the dividend payment is the same as the value of the option immediately after the dividend payment, but at asset value $S(1-d_y)$. Thus, for fixed S the value of the option changes discontinuously across a dividend date. However, (6.7) *is* equivalent to insisting that the option value is continuous in time for any realisation of the asset's random walk.

It is certainly true that the holder of the option does not receive any benefit from the dividend payment, and so the option price must reflect this forgone benefit. The fact that the option price is continuous for each realisation of the asset's random walk, even though the asset value is not, does not mean that the option value is unaffected by dividend payments. The effect of the jump condition (6.6) is felt throughout the life of the option, propagated by the partial differential equation that governs its value.

Finally, note that the delta of an option does change across a dividend date. A corresponding adjustment must be made to any hedged portfolio.

6.2.5 The Call Option with One Dividend Payment

Let us now value a European call with one dividend payment, as above. Recall how we solve in the absence of dividends: because the Black–Scholes equation is backward parabolic, we work backwards from expiry, when we know the value with certainty. When a dividend is paid, this idea is elaborated as follows:

- Solve the Black–Scholes equation back from expiry until just after the dividend date (i.e. until $t = t_d^+$);
- Implement the jump condition (6.7) across $t = t_d$, to find the values at $t = t_d^-$;
- Solve the Black–Scholes equation backwards from $t = t_d^-$, using these values as final data.

In effect, we solve the Black–Scholes equation twice, once for $T > t > t_d$, and once for $t_d > t > 0$ (the present day). The values at $t = t_d^{\pm}$ are linked by (6.7). A very similar structure emerges when we value certain exotic options.

Let us write $C_d(S,t)$ for the value of our call option (in the discussion above, $V(S,t)$ can represent *any* dividend-paying derivative product). Let us also write $C(S,t;E)$ for the value of a vanilla European call option with exercise price E (the other parameters, r, σ and T are understood). For times after the dividend date, our option is identical to a vanilla call:

no more dividends will be paid. Thus,

$$C_d(S,t) = C(S,t;E) \quad \text{for} \quad t_d^+ \le t \le T.$$

Now we use (6.7):

$$\begin{aligned} C_d(S,t_d^-) &= C_d\big(S(1-d_y),t_d^+\big) \\ &= C\big(S(1-d_y),t_d^+;E\big). \end{aligned} \qquad (6.8)$$

At this point, we could use the values we have just calculated in our formula (5.14) for solutions of the Black–Scholes equation. However, there is a short cut. The call option in (6.8) is evaluated not at S, but at $S(1-d_y)$: a uniform scaling of S by $(1-d_y)$. A uniform scaling of this kind leaves the Black–Scholes equation invariant, and so $C\big(S(1-d_y),t;E\big)$ is a solution, which is equal to our option value at $t=t_d^-$, and hence for all times before t_d.

It only remains to identify $C\big(S(1-d_y),t;E\big)$. At expiry, this derivative product has value

$$\begin{aligned} C\big(S(1-d_y),T;E\big) &= \max\big(S(1-d_y)-E,0\big) \\ &= (1-d_y)\max\big(S-E(1-d_y)^{-1},0\big). \end{aligned}$$

Thus, it is the same as $(1-d_y)$ calls *with exercise price* $E(1-d_y)^{-1}$. For times before t_d, our call has value

$$C_d(S,t) = (1-d_y)C\big(S,t;E(1-d_y)^{-1}\big).$$

Note that the effect of the dividend is to decrease the value of the call. This is reasonable because the option holder does not receive the dividend, and the effect of the latter is to decrease S and hence the upside potential of the option.

Technical Point: Continuous and Discrete Dividends Unified. In this Technical Point, we outline a way to unify continuous and discrete dividend payments. For more details, see *Option Pricing.* Suppose the dividend structure is a quite general function of S and t, $D(S,t)$. The constant-yield case above had $D(S,t)=D_0S$, while in the discrete case, $D(S,t)=D_\delta S\delta(t-t_d)$ for some constant D_δ, which we relate to d_y below. As above, the stochastic differential equation (2.1) describing the random walk followed by the asset must be modified for the dividend payment, so that it becomes

$$dS = \sigma S\,dX + \big(\mu S - D(S,t)\big)\,dt. \qquad (6.9)$$

When the dividend payment is discrete, this gives

$$dS = \sigma S\, dX + \big(\mu S - D_\delta S \delta(t - t_d)\big)\, dt.$$

Integrating across the dividend date, we find that

$$\int_{S(t_d^-)}^{S(t_d^+)} \frac{dS}{S} = \int_{t_d^-}^{t_d^+} \sigma\, dX + \int_{t_d^-}^{t_d^+} \mu\, dt - D_\delta \int_{t_d^-}^{t_d^+} \delta(t - t_d)\, dt.$$

Since t_d^- and t_d^+ differ only infinitesimally, the only nonzero term on the right-hand side is the one containing the delta function, and hence we obtain

$$\int_{S(t_d^-)}^{S(t_d^+)} \frac{dS}{S} = \log\left(\frac{S(t_d^+)}{S(t_d^-)}\right) = -D_\delta \int_{t_d^-}^{t_d^+} \delta(t - t_d)\, dt = -D_\delta. \qquad (6.10)$$

Thus, for a discrete payment $D_\delta S \delta(t - t_d)$, the asset is discounted by

$$e^{D_\delta \mathcal{H}(t - t_d)};$$

consequently, $D_\delta = -\log d_y$. (Thus if a company pays out half of the asset price at time t_d, this discretely paid constant dividend yield gives $e^{-D_\delta} = \frac{1}{2}$ (this is d_y), $D_\delta^y = \log 2$.)

For any given realisation the value of the option is continuous, and hence the appropriate jump condition is

$$V\big(S(t_d^+), t_d^+\big) = V\big(S(t_d^-), t_d^-\big)$$

with $S(t_d^+)$ and $S(t_d^-)$ related by (6.10).

6.3 Forward and Futures Contracts

Forward and futures contracts are in some ways easier to value than options. This is because all the risk can be eliminated by a once-and-for-all hedge at the beginning of the contract. As a corollary, they can be valued *independently* of any assumptions about the behaviour of the asset price, provided only that the future course of interest rates can be predicted. Nevertheless, we prefer to discuss them here in the Black–Scholes framework.

We need only analyse the forward contract, since, as stated in Chapter 1, forward and futures prices are the same (under some not too restrictive assumptions). Recall that the forward price is not set at one of a number of fixed values for all contracts on the same asset with the same expiry. Instead, it is determined at the outset, individually for each contract. Suppose that the time at which the contract is agreed is t, and that the asset price at that time is $S(t)$. Denoting the forward price by F, we must find a relationship between $S(t)$ and F that will ensure fair value for both parties to the contract. We assume that interest rates are constant over the duration of the contract.

There are several ways of deriving the forward price. We begin with one based on arbitrage. Consider first the party who is short the contract, and so must deliver the asset at time T. Although he does not know at time t what the asset price will be at time T, this does not matter. He can satisfy his part of the contract by borrowing an amount $S(t)$ when the contract begins, buying the asset, and using the money received at exercise, F, to pay off the loan. Assuming that the risk-free interest rate r is constant, the loan will cost $S(t)e^{r(T-t)}$. The forward price must therefore be given by

$$F = S(t)e^{r(T-t)}. \tag{6.11}$$

If this were not so, there would be a risk-free profit or loss on the transaction, in contradiction to the absence of arbitrage. A similar argument applies to the party who is long the contract, and yields the same price.

Another way of looking at this result is to notice that a long position in the forward contract is equivalent to a long position in a European call option and a short position in a put option, both with the same expiry and exercise price as the forward contract. (This is just a restatement of the put-call parity result (3.2).) Since the forward contract has zero value when it is set up (no money changes hands), the exercise price of the options, E, which is also equal to the forward price F, must be such that $S - Ee^{-r(T-t)} = 0$; this gives (6.11).

Our final interpretation is perhaps the least obvious of the three, but it is a pointer to the way in which we approach more complicated derivative products. It is not always possible to find a simple 'financial' solution, such as the one above, based on the construction of an equivalent portfolio (here, the asset and a loan), and so to build the answer out of products we know how to value. We step back to see that the forward contract is a derivative contract, albeit of a simple form. Therefore, it must satisfy the Black–Scholes equation. The payoff, at time T, is $S - F$; it is easy to find the solution at any earlier time t as

$$S - Fe^{-r(T-t)}.$$

Since the value of the contract is zero when it is initiated, we arrive at (6.11).

We also note that the value of a forward contract changes with time, because S changes. At any time t' between t and T, a party to a forward contract can lock in a profit (or loss) by entering into the equal and opposite contract. The argument above shows that the value then is

$$V(S,t') = S(t') - Fe^{-r(T-t')}; \tag{6.12}$$

when $t' = t$ the value is zero, and when $t' = T$ it is the payoff, $S(T) - F$.

We have so far assumed that the asset in question pays no dividend. If it pays a constant dividend yield D_0, a simple modification to the argument above shows that the forward price is related to the current price by

$$F = S(t)e^{(r-D_0)(T-t)}. \tag{6.13}$$

(The proof of this is requested in the Exercises.) In some cases D_0 may be negative, an example being the cost of holding an asset such as gold, which has to be stored and insured.

6.4 Options on Futures

Many options have as their underlying asset not the cash product but rather the corresponding futures contract, which is often more liquid and involves lower transaction costs. Options on futures therefore have a value that depends on F and t, i.e. of the form $V(F,t)$. Since

$$F = Se^{r(T-t)},$$

we can derive a partial differential equation for $V(F,t)$ from the ordinary Black–Scholes equation (written in terms of S and t) via the change of variable rule. That is, we replace S by $Fe^{-r(T-t)}$ throughout, and we replace

$$\frac{\partial}{\partial S} \quad \text{by} \quad \frac{\partial F}{\partial S}\frac{\partial}{\partial F} = e^{r(T-t)}\frac{\partial}{\partial F}$$

and

$$\frac{\partial}{\partial t} \quad \text{by} \quad \frac{\partial}{\partial t} + \frac{\partial F}{\partial t}\frac{\partial}{\partial F} = \frac{\partial}{\partial t} - rF\frac{\partial}{\partial F}.$$

The result is

$$\frac{\partial V}{\partial t} + \tfrac{1}{2}\sigma^2\frac{\partial^2 V}{\partial F^2} - rV = 0. \tag{6.14}$$

We can, of course, derive equation (6.14) directly, just as we derive the Black–Scholes equation. There is, though, a slight finesse in the argument. We take the usual hedged portfolio $\Pi = V - \Delta F$. Noting that the volatility of F is also σ (see Exercise 9), we have that $dF^2 = \sigma^2 F^2 dt$. Itô's lemma thus gives

$$dV = \left(\frac{\partial V}{\partial t} + \tfrac{1}{2}\sigma^2 F^2 \frac{\partial^2 V}{\partial F^2} \right) dt + \frac{\partial V}{\partial F} dF,$$

and as expected, the choice $\Delta = \partial V / \partial F$ makes the portfolio instantaneously risk-free, so

$$d\Pi = \left(\frac{\partial V}{\partial t} + \tfrac{1}{2}\sigma^2 F^2 \frac{\partial^2 V}{\partial F^2} \right) dt.$$

Now comes the subtlety. We want to equate $d\Pi$ to the risk-free return on the portfolio. What we really mean here is that if at time t we were to set up the portfolio, we should earn the risk-free rate on the money thus used. The cost of setting up the portfolio is just V, since it costs nothing to enter into a futures contract. Therefore, $d\Pi = rV\,dt$, and we obtain equation (6.14). (At the corresponding point in the derivation of the Black–Scholes equation on an ordinary asset, we have to trade in $-\Delta$ of the asset to set up the hedge, and this involves a cash flow.)

Since equation (6.14) is identical to the Black–Scholes equation when the asset pays dividends at a rate r, we can use the results obtained earlier to price specific contracts. For example, the value of a European call is easily found to be

$$C(F,t) = e^{-r(T-t)} \Big(FN(d_1) - EN(d_2) \Big),$$

where d_1 and d_2 are just as in the usual Black–Scholes formulæ (5.14), but with S replaced by F. The valuation of the put and the put-call parity relationship are left as exercises, as is the case when the asset pays dividends.

6.5 Time-dependent parameters in the Black–Scholes equation

In this section we show how to derive explicit formulæ for options with time-varying interest rate and volatility. A model of this kind may be useful to someone who has, for example, strong views about the likely future course of interest rates or volatility. Although $r(t)$ and $\sigma(t)$ vary, we have to assume that we know how they do so. Random variations in either parameter lead to much more complicated **two-factor** models; in

Chapter 18 we describe models for convertible bonds, which are subject to both asset and interest rate risk.

The Black–Scholes equation (3.9) remains valid even if we replace r and σ by $r(t)$ and $\sigma(t)$; the argument goes through in the same way. Furthermore, we can still solve the equation in this far more general situation. Considering the specific cases of calls and puts, the boundary and final conditions for these options remain exactly as before with one exception. Since r appears explicitly in (3.14), the boundary condition at $S = 0$ for the put, we must be more careful in deriving this condition. When r is a function of time the correct condition is

$$P(0,t) = Ee^{-\int_t^T r(\tau)d\tau}.$$

Since the payoff at expiry is certainly E if S is zero, the value of the put option on $S = 0$ the present value of the payoff under non-constant interest rates.

We can now derive explicit formulæ when both r and σ are functions of time.[2] The equation to be solved is

$$\frac{\partial V}{\partial t} + \tfrac{1}{2}\sigma^2(t)S^2\frac{\partial^2 V}{\partial S^2} + r(t)S\frac{\partial V}{\partial S} - r(t)V = 0, \qquad (6.15)$$

where we have shown the dependence on t explicitly. Let us make the following substitutions:

$$\bar{S} = Se^{\alpha(t)},$$

$$\bar{V} = Ve^{\beta(t)},$$

$$\bar{t} = \gamma(t),$$

where α, β and γ are to be chosen carefully so as to eliminate all time-dependent coefficients from (6.15). In these new variables (6.15) becomes

$$\dot{\gamma}(t)\frac{\partial \bar{V}}{\partial \bar{t}} + \tfrac{1}{2}\sigma(t)^2\bar{S}^2\frac{\partial^2 \bar{V}}{\partial \bar{S}^2} + \big(r(t) + \dot{\alpha}(t)\big)\bar{S}\frac{\partial \bar{V}}{\partial \bar{S}} - \big(r(t) + \dot{\beta}(t)\big)\bar{V} = 0, \quad (6.16)$$

where $\dot{} = d/dt$. Now eliminate the coefficients of $\bar{V}$ and $\partial\bar{V}/\partial\bar{S}$ by choosing

$$\alpha(t) = \int_t^T r(\tau)\,d\tau,$$

$$\beta(t) = \int_t^T r(\tau)\,d\tau,$$

[2] Some readers may have already derived these formulæ by a different procedure, in Exercise 5 at the end of Chapter 5.

and remove the remaining time dependence by setting

$$\gamma(t) = \int_t^T \sigma^2(\tau)\,d\tau.$$

With these choices (6.16) becomes

$$\frac{\partial \bar{V}}{\partial \bar{t}} = \tfrac{1}{2}\bar{S}^2 \frac{\partial^2 \bar{V}}{\partial \bar{S}^2} \tag{6.17}$$

and has coefficients which are *independent of time*: this equation contains no reference to r or σ. If $\bar{V}(\bar{S}, \bar{t}\,)$ is any solution of (6.17), then the corresponding solution of (6.16), in original variables, is

$$V = e^{-\beta(t)}\bar{V}\left(Se^{\alpha(t)}, \gamma(t)\right). \tag{6.18}$$

Let us now denote by V_{BS} any solution of the Black–Scholes equation for constant r, σ and zero dividends. In view of the above, this solution can be written in the form

$$V_{BS} = e^{-r(T-t)}\bar{V}_{BS}\left(Se^{-r(T-t)}, \sigma^2(T-t)\right) \tag{6.19}$$

for some function $\bar{V}_{BS}$. By comparing (6.18) and (6.19) we see that to go from an explicit solution of the Black–Scholes equation with constant r and σ and zero dividends we simply perform the following substitutions:

- wherever we see r in the explicit formula replace it by

$$\frac{1}{T-t}\int_t^T r(\tau)\,d\tau;$$

- wherever we see σ^2 in the explicit formula replace it by

$$\frac{1}{T-t}\int_t^T \sigma^2(\tau)\,d\tau.$$

(Note that these formulæ give the average, over the remaining lifetime of the option, of the interest rate and squared volatility respectively.) The reader should check carefully that this procedure gives a value that is a solution of the modified Black–Scholes equation (6.15), and that for a given option such as a call, the payoff condition is satisfied.

Further Reading

- For the original derivation of the Black–Scholes formulæ with time-dependent parameters see Merton (1973).
- In this book we have assumed that dividends are known. For a model with *stochastic* dividends see Geske (1978).
- Gemmill (1992) discusses the practical implications of discrete dividend payments.

Exercises

1. What is the put-call parity relation for options on an asset that pays a constant continuous dividend yield?
2. What is the delta for the call option with continuous and constant dividend yield?
3. Find a transformation that reduces the Black–Scholes equation with a constant continuous dividend yield to the diffusion equation. What is the transformed payoff for a call? How many dimensionless parameters are there in the problem?
4. Show that the value of a European call option on an asset that pays a constant continuous dividend yield lies below the payoff for large enough values of S. Show also that the call on an asset with dividends is less valuable than the call on an asset without dividends.
5. Calculate the value of a put option for both continuous and discrete dividend yields (one payment). What is the put-call parity relation in the latter case? Do the dividends increase or decrease the value of the put? Why?
6. Calculate the value of a call on an asset that pays out *two* dividends during the lifetime of the option.
7. Another model for dividend structures is to assume that the dividend payment will be a fixed amount D paid at time t_d. Work out the jump condition for a derivative product, and calculate the value of a call and put. What possible disadvantages might this model have?
8. Suppose that a forward contract had the additional condition that a premium Z had to be paid on entering into the contract. How would the forward price be affected?
9. What is the random walk followed by the futures price F?

10. Derive the put-call parity result for the forward/futures price in the form

$$C - P = (F - E)e^{-r(T-t)}.$$

What is the corresponding version when the asset pays a constant continuous dividend yield?

11. What is the forward price for an asset that pays a single dividend $d_y S(t_d)$ at time t_d?

12. Analyse the **range forward** contract, which has the following features. There are two exercise prices, E_1 and E_2, with $E_1 < E_2$. The holder of a long position must purchase the asset for E_1 if at expiry $S < E_1$, for S if $E_1 \le S \le E_2$, and for E_2 if $S > E_2$. The exercise prices are to be chosen so that the initial cost is zero.

13. What is the model for derivative products on an asset that pays a time-varying dividend yield $D(t)S$? Show how to incorporate this variation into the time-dependent version of the Black–Scholes model described in Section 6.5; how are the functions $\alpha(t)$, $\beta(t)$ and $\gamma(t)$ modified?

7 American Options

7.1 Introduction

We recall from Section 1.5 that an American option has the additional feature that exercise is permitted at any time during the life of the option. (Of course, this relies on the assumption that there is a well-defined payoff for early exercise.) The explicit formulæ quoted in Chapter 5, which are valid for European options where early exercise is not permitted, do not necessarily give the value for American options. In fact, since the American option gives its holder greater rights than the European option, via the right of early exercise, potentially it has a higher value. The following arbitrage argument shows how this can happen.

Figure 7.1 shows that before expiry there is a large range of asset values S for which the value of a European put option is less than its intrinsic value (the payoff function). Suppose that S lies in this range, so that $P(S,t) < \max(E-S,0)$, and consider the effect of exercising the option. There is an obvious arbitrage opportunity: we can buy the asset in the market for S, at the same time buying the option for P; if we immediately exercise the option by selling the asset for E, we thereby make a risk-free profit of $E - P - S$. Of course, such an opportunity would not last long before the value of the option was pushed up by the demand of arbitragers. We conclude that when early exercise is permitted we must impose the constraint

$$V(S,t) \geq \max(S - E, 0). \tag{7.1}$$

American and European put options must therefore have different values.

A second example of an American option whose value differs from that of its European equivalent is a call option on a dividend-paying

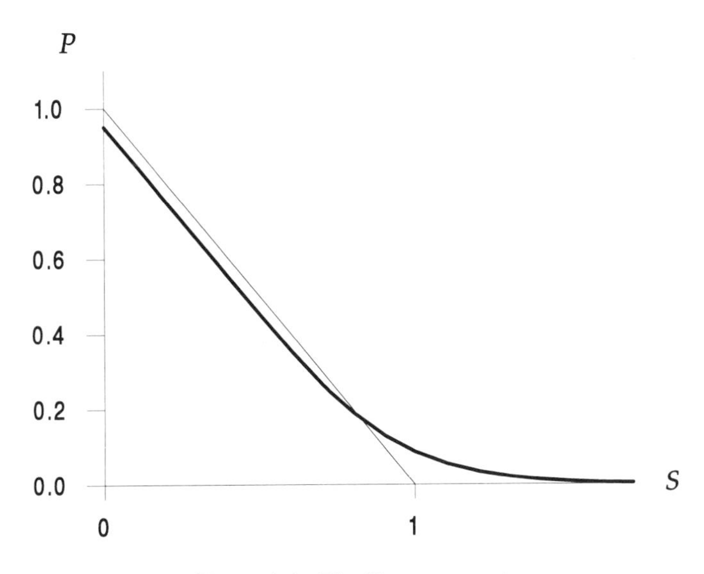

Figure 7.1. The European put

asset. Recall from equation (6.4) that for large values of S, the dominant behaviour of the European option is

$$C(S,t) \sim Se^{-D_0(T-t)}.$$

If $D_0 > 0$, this certainly lies below the payoff $\max(S-E,0)$ for large S, and an arbitrage argument as above shows that the American version of this option must also be more valuable than the European version, since it must satisfy the constraint

$$C(S,t) \geq \max(S-E,0).$$

In both of these cases, there must be some values of S for which it is optimal from the holder's point of view to exercise the American option. If this were not so, then the option would have the European value, since the Black–Scholes equation would hold for all S: we have seen that this is not the case. The valuation of American options is therefore more complicated, since at each time we have to determine not only the option value, but also, for each value of S, whether or not it should be exercised. This is what is known as a **free boundary problem**. Typically at each time t there is a particular value of S which marks the boundary between two regions: to one side one should hold the option and to the other side one should exercise it. (There may be several such values; for the moment we suppose that there is just one.) We denote this value, which in general varies with time, by $S_f(t)$, and refer to it as the **optimal exercise price**. Since we do not know S_f *a priori* we are lacking one piece of information compared with the corresponding European valuation problem. With the European option we know which boundary conditions to apply and, equally importantly, *where* to apply them. With the American problem we do not know *a priori* where to apply boundary conditions; the unknown boundary $S_f(t)$ is for this reason called a **free boundary**. This situation is common to many financial and physical problems; as a canonical example and an aid to intuition, as well as an introduction to the mathematical and, later, numerical techniques available, we mention the **obstacle problem**.

7.2 The Obstacle Problem

At its simplest, an obstacle problem arises when an elastic string is held fixed at two ends, A and B, and passes over a smooth object which protrudes between the two ends (see Figure 7.2). Again, we do not know *a priori* the region of contact between the string and the obstacle, only that either the string is in contact with the obstacle, in which case its position is known, or it must satisfy an equation of motion, which, in this case, says that it must be straight. Beyond this, the string must satisfy two constraints. The first simply says that the string must lie above or on the obstacle; combined with the equation of motion, the curvature of the string must be negative or zero. Another interpretation of this is that the obstacle can never exert a negative force on the string: it can push but not pull. The second constraint on the string is that its slope must be continuous. This is obvious except at points where the string first loses contact with the obstacle, and there it is justified by a local force balance: a lateral force is needed to create a kink in the string, and there is none. In summary,

- the string must be above or on the obstacle;
- the string must have negative or zero curvature;
- the string must be continuous;
- the string slope must be continuous.

Under these constraints, the solution to the obstacle problem can be shown to be unique. The string and its slope are continuous, but in general the curvature of the string, and hence its second derivative, has discontinuities.

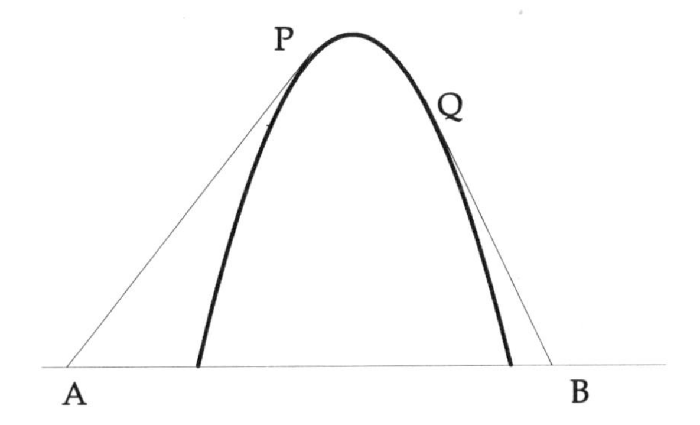

Figure 7.2 The classical obstacle problem: the string is held fixed at A and B and must pass smoothly over the obstacle in between.

7.3 American Options as Free Boundary Problems

An American option valuation problem can also be shown to be uniquely specified by a set of constraints, very similar to those just given for the obstacle problem. They are:

- the option value must be greater than or equal to the payoff function;
- the Black–Scholes equation is replaced by an inequality (this is made precise shortly);
- the option value must be a continuous function of S;
- the option delta (its slope) must be continuous.

The first of these constraints says that the arbitrage profit obtainable from early exercise must be less than or equal to zero. It does not mean that early exercise should never occur, merely that arbitrage opportunities should not. Thus, either the option value is the same as the payoff function, and the option should be exercised, or, where it exceeds the payoff, it satisfies the appropriate Black–Scholes equation. It turns out that these two statements can be combined into one inequality for the Black–Scholes equation, which is our second constraint above. There are some interesting features to this inequality, and we return to it briefly below.

The third constraint, that the option value is continuous, follows from simple arbitrage. If there were a discontinuity in the option value as a function of S, and if this discontinuity persisted for more than an infinitesimal time, a portfolio of options only would make a risk-free profit with probability 1 should the asset price ever reach the value at which the discontinuity occurred.[1]

Just as in the obstacle problem, we do not know the position of S_f, and we must impose *two* conditions at S_f if the option value is to be uniquely determined. This is one more than if S_f were specified. The second condition at S_f, our fourth constraint above, is that the option delta must also be continuous there. Its derivation is rather more delicate, and we only give an informal financially based argument, for the specific case of the American put.

7.4 The American Put

Consider the American put option, with value $P(S,t)$. We have already argued that this option has an exercise boundary $S = S_f(t)$, where the option should be exercised if $S < S_f(t)$ and held otherwise. Assuming that $S_f(t) < E$, the slope of the payoff function $\max(E - S, 0)$ at the contact point is -1. There are three possibilities[2] for the slope (delta) of the option, $\partial P/\partial S$, at $S = S_f(t)$:

- $\partial P/\partial S < -1$;
- $\partial P/\partial S > -1$;
- $\partial P/\partial S = -1$.

We show that the first two are incorrect.

Suppose first that $\partial P/\partial S < -1$. Then as S increases from $S_f(t)$, $P(S,t)$ drops below the payoff $\max(E - S, 0)$, since its slope is more negative; see Figure 7.3(a). This contradicts our earlier arbitrage bound $P(S,t) \geq \max(E - S, 0)$, and so is impossible.

Now suppose that $\partial P/\partial S > -1$, as in Figure 7.3(b). In this case, we argue that an option value with this slope would be sub-optimal for the holder, in the sense that it does not give the option its maximum value consistent with the Black–Scholes risk-free hedging strategy and the arbitrage constraint $P(S,t) \geq \max(E-S, 0)$. In order to see this, we must discuss the strategy adopted by the holder. There are two aspects to consider. One is the day-to-day arbitrage-based hedging strategy which,

[1] This result does *not* imply that there is a blanket prohibition of discontinuous option prices, caused for example by an instantaneous change in the terms of the contract such as the imposition of a constraint by a change from European to American. Indeed, such discontinuities, or jumps, play an important part in later chapters.

[2] A fourth is that $\partial P/\partial S$ does not exist at $S = S_f(t)$. We assume, as can be shown to be the case, that it does.

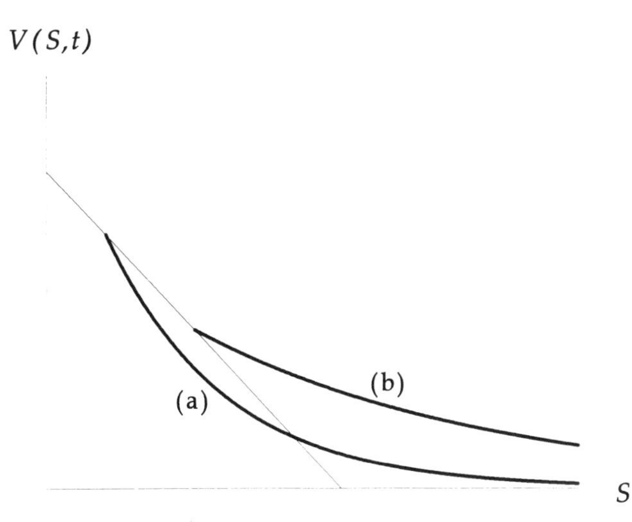

Figure 7.3. Exercise price (a) too low (b) too high.

as above, leads to the Black–Scholes equation. The other is the **exercise strategy**: the holder must decide, in principle, how far S should fall before he would exercise the option. The basis of this decision is, naturally enough, that the chosen strategy should maximise an appropriate measure of the value of the option to its holder.[3] Because the option satisfies a partial differential equation with $P(S_f(t), t) = E - S_f(t)$ as one of the boundary conditions, the choice of $S_f(t)$ affects the value of $P(S, t)$ for *all* larger values of S. Clearly the case of Figure 7.3(a) corresponds to too low a value of $S_f(t)$, and an arbitrage profit is possible for S just above $S_f(t)$. Conversely, if $\partial P/\partial S > -1$ at $S = S_f(t)$, the value of the option near $S = S_f(t)$ can be increased by choosing a smaller value for S_f: the exercise value then moves up the payoff curve and $\partial P/\partial S$ decreases. The option is thus again misvalued. In fact, the increase in P is passed on by the partial differential equation to all values of S greater than S_f, and by decreasing S_f we arrive at the crossover point between our two incorrect possibilities, which simultaneously maximises the benefit to the holder and avoids arbitrage. This yields the correct free boundary condition $\partial P/\partial S = -1$ at $S = S_f(t)$.

[3] This choice also minimises the benefit to the writer, but since the holder can close the contract by exercising and the writer cannot, the latter's point of view is not relevant to this argument. Of course, the writer requires a greater premium in recompense for the one-sided nature of the contract.

We must stress that the argument just given is not a rigorous formal derivation of the second free boundary condition. Such a derivation might be couched in the language of stochastic control and optimal stopping problems, or of game theory; both are beyond the scope of this book. Suffice it to say that the correct formulation of a rational operator's strategy when holding an American option can be shown to lead to the condition that the option value meets the payoff function smoothly, as long as the latter is smooth too.

Finally, we return to the second constraint above, the 'inequality' satisfied by the Black–Scholes operator. Recall that the Black–Scholes partial differential equation follows from an arbitrage argument. This argument is only partially valid for American options, but the intimate relationship between arbitrage and the Black–Scholes operator persists; the former now yields an inequality (rather than an equation) for the latter.

We set up the delta-hedged portfolio as before, with exactly the same choice for the delta. However, in the American case it is not necessarily possible for the option to be held both long and short, since there are times when it is optimal to exercise the option. Thus, the writer of an option may be exercised against. The simple arbitrage argument used for the European option no longer leads to a unique value for the return on the portfolio, only to an inequality. We can only say that the return from the portfolio cannot be greater than the return from a bank deposit. For an American put, this gives

$$\frac{\partial P}{\partial t} + \tfrac{1}{2}\sigma^2 S^2 \frac{\partial^2 P}{\partial S^2} + rS\frac{\partial P}{\partial S} - rP \leq 0. \tag{7.2}$$

The *inequality* here would be an *equality* for a European option. When it is optimal to hold the option the equality, i.e. the Black–Scholes equation, is valid and the constraint (7.1) must be satisfied. Otherwise, it is optimal to exercise the option, and only the inequality in (7.2) holds and the equality in (7.1) is satisfied—the payoff, or obstacle, is the solution. It is easy to verify that this is so: when $P = E - S$, for $S < E$, substitution into (7.2) gives

$$\frac{\partial P}{\partial t} + \tfrac{1}{2}\sigma^2 S^2 \frac{\partial^2 P}{\partial S^2} + rS\frac{\partial P}{\partial S} - rP = -rE < 0.$$

In summary, the American put problem is written as a free boundary problem as follows. For each time t, we must divide the S axis into two distinct regions. The first, $0 \leq S < S_f(t)$, is where early exercise is

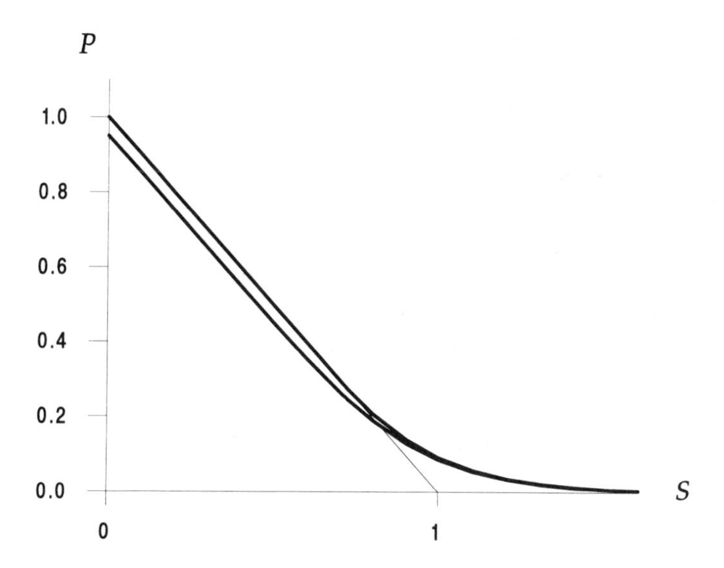

Figure 7.4 The values of European and American put options as functions of S; $r = 0.1$, $\sigma = 0.4$, $E = 1$ and six months to expiry. The upper curve is the value of the American put, which joins smoothly onto the payoff function (also shown).

optimal and

$$P = E - S, \quad \frac{\partial P}{\partial t} + \tfrac{1}{2}\sigma^2 S^2 \frac{\partial^2 P}{\partial S^2} + rS\frac{\partial P}{\partial S} - rP < 0.$$

In the other region, $S_f(t) < S < \infty$, early exercise is not optimal and

$$P > E - S, \quad \frac{\partial P}{\partial t} + \tfrac{1}{2}\sigma^2 S^2 \frac{\partial^2 P}{\partial S^2} + rS\frac{\partial P}{\partial S} - rP = 0.$$

The boundary conditions at $S = S_f(t)$ are that P and its slope (delta) are continuous:

$$P\left(S_f(t), t\right) = \max\left(E - S_f(t), 0\right), \quad \frac{\partial P}{\partial S}\left(S_f(t), t\right) = -1. \qquad (7.3)$$

We can think of these as being one boundary condition to determine the option value on the free boundary, and the other to determine the location of the free boundary. It is very important to realise that the condition

$$\frac{\partial P}{\partial S}\left(S_f(t), t\right) = -1$$

is *not* implied by the fact that $P(S_f(t), t) = E - S_f(t)$. Since we do not know *a priori* where $S_f(t)$ is, we need an extra condition to determine it.

Arbitrage arguments show that the gradient of P should be continuous, and this gives us the extra condition we require.

In Figure 7.4 we compare the values of European and American put options at six months before expiry with $\sigma = 0.4$ and $r = 0.1$. The former is given by the explicit formula (3.18) and the latter has been calculated numerically by the methods of Chapter 9.

Technical Point: Free Boundary Conditions.
We emphasise also that both the boundary conditions for the American put are based on financial reasoning, namely arbitrage. Many other candidates are equally possible from the purely mathematical point of view; although it would not be a model for option pricing, we would also get a well-posed free boundary problem if we imposed a condition such as

$$\frac{\partial P}{\partial S}\big(S_f(t), t\big) = 0 \quad \text{if} \quad P\big(S_f(t), t\big) = E - S_f(t),$$

or one such as

$$\frac{\partial P}{\partial S}\big(S_f(t), t\big) = -\frac{dS_f}{dt}.$$

This latter condition is, in fact, the proper free boundary condition for the Stefan model of melting ice. It is, of course, a totally inappropriate condition for American puts.

7.5 Other American Options

The arguments we have just given for the American put apply, with appropriate modifications, to any vanilla option or combination of options with payoff $\Lambda(S)$, or even $\Lambda(S,t)$. Including a constant dividend yield, the option value $V(S,t)$ satisfies the Black–Scholes inequality

$$\frac{\partial V}{\partial t} + \tfrac{1}{2}\sigma^2 S^2 \frac{\partial^2 V}{\partial S^2} + (r - D_0)S\frac{\partial V}{\partial S} - rV \le 0.$$

Where exercise is optimal, $V(S,t) = \Lambda(S)$ and the inequality is strict ($<$ rather than $\le$). Otherwise $V(S,t) > \Lambda(S)$, and the inequality is an equality. At the free boundary (or boundaries, for there may be several of them if the payoff is complicated enough), both V and $\partial V/\partial S$ must be continuous. The specification of the problem is completed by the terminal condition

$$V(S,T) = \Lambda(S),$$

together with appropriate conditions at infinity. We see other examples of such partial differential inequalities in later chapters on exotic options.

Technical Point: Options with Discontinuous Payoffs.
The condition that the Δ for an American option must be continuous assumes that the payoff function itself has a continuous slope. That is, it is possible for the option value to meet the payoff tangentially only if the payoff has a well-defined tangent at the point of contact.

As an example, consider an American cash-or-nothing call option with payoff given by

$$V(S,T) = \begin{cases} 0 & S < E \\ B & S \geq E. \end{cases}$$

The payoff is discontinuous. The option value is continuous except at expiry, but the Δ is discontinuous at $S = E$. It is clear that the optimal exercise boundary is always at $S = E$; there is no gain to be made from holding such an option once S has reached the exercise price. Indeed, interest on the payoff is lost if the option is held after S has reached E. Thus, there is no point in hedging for $S > E$; looked at another way, $\Delta = 0$ for $S > E$. Clearly $\Delta > 0$ for $S < E$.

Mathematically, we find that we have two boundary conditions, namely $V(0,t) = 0$, $V(E,t) = B$ and a payoff condition $V(S,T) = 0$ for $0 \leq S \leq E$. There is no point in considering values of $S > E$, since the option would have been exercised. Unlike the usual American option, where 'spatial' boundary conditions are applied at an unknown value of S, and so an extra condition is needed, both the spatial conditions here are at *known* values of S. These three conditions therefore give a unique solution of the Black–Scholes equation. This option is one of the few American options to have a useful explicit solution (see Exercise 6 of this chapter).

This option also illustrates very well the idea that the exercise strategy for an American option should maximise its value to the holder. It is particularly clear here that the choice $S_f(t) = E$ gives the largest values of $V(S,t)$ for all $S < E$, as illustrated in Figure 7.5.

7.6 Linear Complementarity Problems

It is clear from the discussion above that the mathematical analysis of American options is more complicated than that of European options. It is almost always impossible to find a useful explicit solution to any given free boundary problem, and so a primary aim is to construct efficient

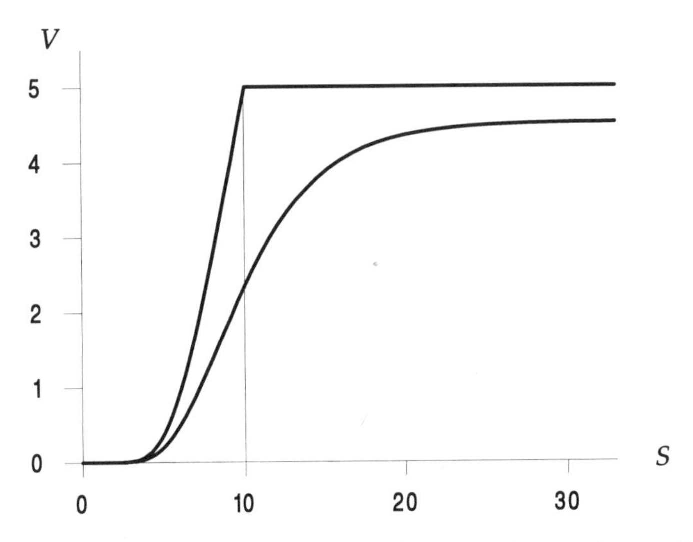

Figure 7.5 European and American values of a cash-or-nothing call with $E = 10$, $B = 5$, $r = 0.1$, $\sigma = 0.4$, $D = 0.02$ and one year to expiry.

and robust numerical methods for their computation. This means that we need a theoretical framework within which to analyse free boundary problems in fairly general terms.

Our starting point is the idea that, since it is difficult to deal with free boundaries, it is worth the effort of attempting to reformulate the problem in such a way as to eliminate any explicit dependence on the free boundary. The free boundary does not then interfere with the solution process, and it can be recovered from the solution *after* the latter has been found. We start by considering a simple example of such a reformulation, called a linear complementarity problem, in the context of the obstacle problem. We then apply the lessons learnt from the obstacle problem to more complicated American options. These problems too have linear complementarity formulations which lead to efficient and accurate numerical solution schemes with the desirable property of not requiring explicit tracking of the free boundary (the same ideas apply equally well to complicated exotic options). These methods are discussed in detail in Chapter 9. The linear complementarity approach also leads to the idea of a variational inequality, and thence to existence and uniqueness proofs for the solution.

7.6.1 The Obstacle Problem

Consider the obstacle problem described in Section 7.2, in which we take the ends of the string to be at $x = \pm 1$ and write $u(x)$ for the string displacement and $f(x)$ for the height of the obstacle, both for $-1 \leq x \leq 1$. We assume that $f(\pm 1) < 0$, and that $f(x) > 0$ at some points $-1 < x < 1$, so that there definitely is a contact region. We also assume, at least initially, that $f'' < 0$, where $' = d/dx$, thereby guaranteeing that there is only one contact region. The free boundary is then the set of points, marked as P $(x = x_P)$ and Q $(x = x_Q)$ in Figure 7.2, where the string first meets the obstacle. These are *a priori* unknown, and have to be determined as part of the solution.

In the contact region, $u = f$, while where the string is not in contact with the obstacle it is straight, so $u'' = 0$. Normally, one would need just two boundary conditions to determine the straight portions of the string uniquely, and the values of u at the two ends of each straight portion would certainly do; indeed, we do have $u(-1) = 0$, $u(x_P) = f(x_P)$ and similar conditions for the other straight portion. However, because P and Q are unknown, we need two more boundary conditions than usual in order to determine these points, and here a physical argument based on a force balance shows that at points such as P and Q, u' must be continuous as well as u. As a free boundary problem we can write the particular example given in Figure 7.2 as the problem of finding $u(x)$ *and* the points P, Q such that

$$
\begin{array}{rcc}
 & u(-1) = 0, & \\
u'' = 0, & & -1 < x < x_P, \\
u(x_P) = f(x_P), & & u'(x_P) = f'(x_P), \\
u(x) = f(x), & & x_P < x < x_Q, \\
u(x_Q) = f(x_Q), & & u'(x_Q) = f'(x_Q), \\
u'' = 0, & & x_Q < x < 1, \\
 & u(1) = 0. &
\end{array}
\tag{7.4}
$$

Given any particular $f(x)$ with the same general shape as in Figure 7.2 it is straightforward in principle to show that $u(x)$, P and Q are uniquely determined by this problem, and to find them. The procedure is tedious, and for all but specially simple f, P and Q must be determined numerically as solutions of an algebraic or transcendental equation. The details are even more complicated when f'' is not always less than or equal to zero, because then multiple contact regions can occur, but again, in principle, it can be done.

An alternative approach to the problem is to note that the string either lies above the obstacle, $u > f$, in which case it is straight, $u'' = 0$, or is in contact with the obstacle, $u = f$, in which case $u'' = f'' < 0$. This means that we can write the problem as what is called a **linear complementarity problem:**[4]

$$u'' \cdot (u - f) = 0, \quad -u'' \geq 0, \quad (u - f) \geq 0, \tag{7.5}$$

subject to the conditions that

$$u(-1) = u(1) = 0, \quad u, u' \text{ are continuous.} \tag{7.6}$$

This statement of the problem has a tremendous advantage over the free boundary version (7.4): there is no explicit mention of the free boundary points A and B. They are still present, but only implicitly via the constraint $u \geq f$. If we can devise an algorithm to solve the constrained problem, we just have to look at the resulting values of $u - f$; the free boundaries are where this function switches from being zero to nonzero. One such algorithm is the Projected SOR algorithm described in Chapter 9 in the context of American options; this is an iterative procedure which, starting with an initial guess for u that is certainly above f, produces a sequence of ever more accurate approximations to the true solution. The constraint is simply implemented, for if values of u less than f are generated, they are simply reset to equal f.

It is beyond the scope of this book to prove that the linear complementarity formulation is equivalent to the free boundary formulation (the hard part being to prove that any solution of the former is also a solution of the latter), nor do we show that there is a unique solution to the former. The proofs use techniques of functional analysis, in particular the theory of variational inequalities, but the basic idea is simply minimisation of the appropriate energy functional over the convex space of all suitably smooth functions $v(x)$ that satisfy the constraint $v \geq f$.

7.6.2 A Linear Complementarity Problem for the American Put Option

We now extend the analogy between the obstacle problem and the Black–Scholes formulation of the free boundary problem for an American put by

[4] In general, a problem of the form

$$\mathcal{A}\mathcal{B} = 0, \quad \mathcal{A} \geq 0, \quad \mathcal{B} \geq 0,$$

is called a complementarity problem, and in this example the factors $\mathcal{A} = u''$ and $\mathcal{B} = u - f$ are both linear in u and f.

showing that the latter can also be reduced to a linear complementarity problem. In fact, the 'spatial' parts of the two problems are almost identical, and the major difference is that the put gives an evolution problem, in contrast to the static obstacle problem. The analogy also carries over to the numerical methods for the two problems: the 'spatial' part of the put problem is solved by the same Projected SOR algorithm, while the 'temporal', or evolution, part is used to time-step the solution forwards.

We first transform the American put problem from the original (S, t) variables to (x, τ) as before. We see in later chapters that this is in some ways better from the numerical point of view, and some of the manipulations are easier. These transformations were given in Chapter 5; the only difference now is that there is an optimal exercise boundary. In original variables this was $S = S_f(t)$, and we write it as $x = x_f(\tau)$; note that because $S_f(t) < E$, $x_f(\tau) < 0$. Also, the payoff function $\max(E - S, 0)$ becomes the function

$$g(x,\tau) = e^{\frac{1}{2}(k+1)^2\tau} \max\left(e^{\frac{1}{2}(k-1)x} - e^{\frac{1}{2}(k+1)x}, 0\right). \tag{7.7}$$

Thus, we have

$$\begin{aligned} \frac{\partial u}{\partial \tau} = \frac{\partial^2 u}{\partial x^2} \quad &\text{for } x > x_f(\tau), \\ u(x,\tau) = g(x,\tau) \quad &\text{for } x \le x_f(\tau), \end{aligned} \tag{7.8}$$

with the initial condition

$$u(x,0) = g(x,0) = \max\left(e^{\frac{1}{2}(k-1)x} - e^{\frac{1}{2}(k+1)x}, 0\right) \tag{7.9}$$

and the asymptotic behaviour

$$\lim_{x\to\infty} u(x,\tau) = 0. \tag{7.10}$$

(As $x \to -\infty$, we are in the region where early exercise is optimal, and so $u = g$.) We also have the crucial constraint

$$u(x,\tau) \ge e^{\frac{1}{2}(k+1)^2\tau} \max\left(e^{\frac{1}{2}(k-1)x} - e^{\frac{1}{2}(k+1)x}, 0\right) \tag{7.11}$$

and the conditions that u and $\partial u/\partial x$ are continuous at $x = x_f(t)$, all of which follow from the corresponding conditions in the original problem.

In order to avoid technical complications, let us accept that, since we are going to have to restrict any numerical scheme to a finite mesh, we may as well restrict the problem to a finite interval. That is, we consider the problem (7.8)–(7.11) only for x in the interval $-x^- < x < x^+$, where x^+ and x^- are large. This means that we impose the boundary conditions

$$u(x^+,\tau) = 0, \quad u(-x^-,\tau) = g(-x^-,\tau). \tag{7.12}$$

In financial terms, we assume that we can replace the exact boundary conditions by the approximations that for small values of S, $P = E - S$, while for large values, $P = 0$.

The fact that both the obstacle problem and the American put problem satisfy constraints suggests that the latter might also have a linear complementarity formulation, and this is indeed the case. The option problem is very similar to the obstacle problem, but with an obstacle which is time-dependent, that is, the transformed payoff function $g(x,\tau)$.

We can write (7.8)–(7.11) in the linear complementarity form

$$\begin{gathered}\left(\frac{\partial u}{\partial \tau} - \frac{\partial^2 u}{\partial x^2}\right) \cdot \Big(u(x,\tau) - g(x,\tau)\Big) = 0, \\ \left(\frac{\partial u}{\partial \tau} - \frac{\partial^2 u}{\partial x^2}\right) \geq 0, \qquad \Big(u(x,\tau) - g(x,\tau)\Big) \geq 0,\end{gathered} \tag{7.13}$$

with the initial and boundary conditions (7.9) and (7.12), namely

$$u(x,0) = g(x,0),$$

$$u(-x^-,\tau) = g(-x^-,\tau), \quad u(x^+,\tau) = g(x^+,\tau) = 0,$$

and the conditions that

$$u(x,\tau) \ \text{ and } \ \frac{\partial u}{\partial x}(x,\tau) \ \text{ are continuous.} \tag{7.14}$$

The two possibilities in this formulation correspond to situations in which it is optimal to exercise the option ($u = g$) and those in which it is not ($u > g$).

Once again, the great advantage of this formulation is that the free boundary (or boundaries) need not be tracked explicitly. As with the obstacle problem, a considerable amount of work is necessary to prove that the linear complementarity formulation is equivalent to the original free boundary problem, and that there is a unique solution. Again, the techniques used are those of functional analysis and parabolic variational inequalities, and more details can be found in the books referred to above.

7.7 The American Call with Dividends

(This section may be omitted at a first reading without loss of continuity.)

We now consider some analytical aspects of the model for an American call option on a dividend-paying asset, introduced in Section 6.2. Recall that the value $C(S,t)$ of the call satisfies

$$\frac{\partial C}{\partial t} + \tfrac{1}{2}\sigma^2 S^2 \frac{\partial^2 C}{\partial S^2} + (r - D_0)S\frac{\partial C}{\partial S} - rC = 0 \tag{7.15}$$

so long as exercise is not optimal. The payoff condition is

$$C(S,T) = \max(S - E, 0), \tag{7.16}$$

and because the option can be exercised at any time, we always have

$$C(S,t) \geq \max(S - E, 0). \tag{7.17}$$

If there is an optimal exercise boundary $S = S_f(t)$ (and we shortly see that there is), then at $S = S_f(t)$,

$$C\big(S_f(t), t\big) = S_f(t) - E, \quad \frac{\partial C}{\partial S}\big(S_f(t), t\big) = 1. \tag{7.18}$$

If an optimal exercise boundary does exist, then (7.15) is valid only while $C(S,t) > \max(S - E, 0)$, since a direct calculation shows that $\max(S - E, 0)$ is not a solution of the Black–Scholes equation (7.15). Again, (7.15) can be replaced by the inequality

$$\frac{\partial C}{\partial t} + \tfrac{1}{2}\sigma^2 S^2 \frac{\partial^2 C}{\partial S^2} + (r - D_0)S\frac{\partial C}{\partial S} - rC \leq 0,$$

in which equality holds only when $C(S,t) > \max(S - E, 0)$. As in the case of the American put, the financial reason for this is that, if early exercise is optimal, then it is so because the option would be less valuable if it were held than if it were exercised immediately and the funds deposited in a bank.

7.7.1 General Results on American Call Options

In what follows we shall assume that the interest rate and dividend yield satisfy $r > D_0 > 0$. As for the European call, it is convenient to make (7.15)–(7.18) dimensionless and to reduce (7.15) to a constant coefficient

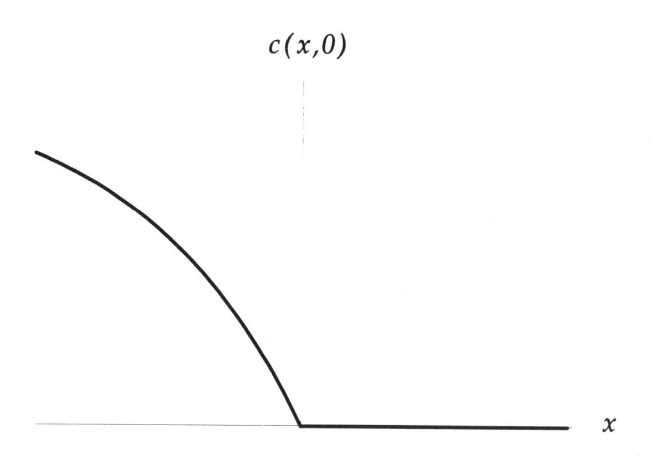

Figure 7.6. $c(x,0)$ for the American call problem.

and forward equation. It also happens to be helpful here to subtract off the payoff $S-E$ from the call value $C(S,t)$. We therefore put

$$S = Ee^x, \quad t = T - \tau/\tfrac{1}{2}\sigma^2, \quad C(S,t) = S - E + Ec(x,\tau),$$

and the result is

$$\frac{\partial c}{\partial \tau} = \frac{\partial^2 c}{\partial x^2} + (k'-1)\frac{\partial c}{\partial x} - kc + f(x) \tag{7.19}$$

for $-\infty < x < \infty$, $\tau > 0$, with

$$c(x,0) = \max(1-e^x, 0) = \begin{cases} 1-e^x & x < 0 \\ 0 & x \geq 0; \end{cases} \tag{7.20}$$

$f(x)$ is defined in (7.22). The graph of the function $c(x,0)$ is sketched in Figure 7.6. The two dimensionless parameters k (which also appeared in the European call solution earlier) and k' are given by

$$k = r/\tfrac{1}{2}\sigma^2, \quad k' = (r - D_0)/\tfrac{1}{2}\sigma^2. \tag{7.21}$$

The function f is given by

$$f(x) = (k'-k)e^x + k. \tag{7.22}$$

Since $r > D_0 > 0$, it follows that $k > k' > 0$.

Assume for the moment that a free boundary does exist, and as before call it $x = x_f(t)$ (in original variables, $S = S_f(t)$). Then at this

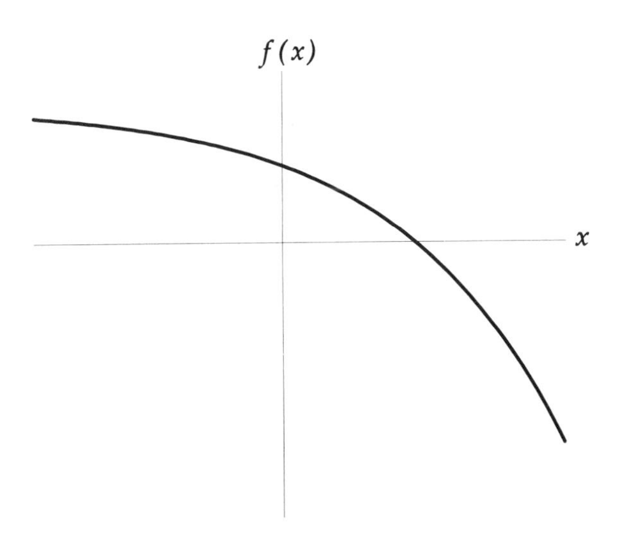

Figure 7.7. The consumption/replenishment term $f(x)$.

boundary we have

$$c\big(x_f(\tau),\tau\big) = \frac{\partial c}{\partial x}\big(x_f(\tau),\tau\big) = 0.$$

Note that the boundary conditions on the free boundary have been simplified; this is the reason for subtracting the payoff function from the call value. The constraint $C \geq \max(S - E, 0)$ becomes

$$c \geq \max(1 - e^x, 0).$$

The behaviour of $f(x)$ is crucial to the behaviour of the free boundary. Indeed, when $f(x)$ is given by (7.22) the existence of this term implies the existence of the optimal exercise boundary. To see this, let us consider further its graph, sketched in Figure 7.7. We see that $f(x)$ is positive for $x < x_0$ where

$$x_0 = \log\left(k/(k - k')\right) = \log(r/D_0) > 0,$$

and negative for $x \geq x_0$.

We now look at what would happen if there were no constraint and thus no free boundary. Consider the initial data $c(x, 0)$ for positive values of x. For $x > 0$, $c(x,0) = \partial c(x,0)/\partial x = \partial^2 c(x,0)/\partial x^2 = 0$. Thus, from (7.19) at expiry we have

$$\frac{\partial c}{\partial \tau} = f(x).$$

For $0 < x < x_0$, $f(x) > 0$ and c immediately becomes positive. For $x > x_0$, $f(x) < 0$ and c immediately becomes negative. Unfortunately, the latter does not satisfy the constraint, which for $x > 0$ requires that c remain positive. If we hold the option in $x > x_0$ the constraint is violated and the option value has fallen below its intrinsic value. This is impossible for an American style option, and hence we deduce that there must be an optimal exercise boundary.

Moreover, it is clear from this argument where the optimal exercise boundary $x_f(\tau)$ must start from. We must have $x_f(0^+) = x_0$, since this is the only point that is consistent with $c\big(x_f(0^+), 0^+\big) = 0$. Financially, this corresponds to

$$S_f(T) = \frac{rE}{D_0},$$

and is independent of σ. Thus, immediately before expiry, the option should be exercised at asset values such that the return on the asset, D_0S, exceeds the interest rate return on the exercise price, rE. At expiry, of course, it will be exercised if $S > E$; here the optimal exercise boundary jumps discontinuously at $t = T$. Note also that if $D_0 = 0$, $x_f = \infty$ ($S_f(T) = \infty$) and there is no free boundary: without dividends we recover the well-known result that it is always optimal to hold an American call to expiry.

We also point out that the point $S = rE/D$ is the value of S at which

$$\mathcal{L}_{BS}\big(\max(S - E, 0)\big) = 0.$$

Technical Point: Physical Interpretation.
Equation (7.19) is a little more complicated than the diffusion equation; there are three extra terms, $(k' - 1)\partial c/\partial x$, $-kc$ and $f(x)$. The first of these can be interpreted as a **convection** term, the second as a **reaction** term and the third, $f(x)$, can be interpreted as a **consumption** term where $f(x) < 0$ or a **replenishment** term where $f(x) > 0$. In order to understand the meaning of these terms and their effect on $c(x, t)$, let us consider combinations of these with other terms in (7.19).

Consider the term $(k - 1)\partial c/\partial x$. Physically, this represents convection (financially, drift). This may be seen by balancing it against the $\partial c/\partial \tau$ on the left-hand side of the partial differential equation and, for the moment, dropping all the other terms. This balance gives the first order hyperbolic equation $\partial c/\partial \tau = (k' - 1)\partial c/\partial x$. This equation may be solved by the method of characteristics to yield $c(x, \tau) = F\big(x + (k' - 1)\tau\big)$ for some function F. Since c is constant along the lines $x + (k' - 1)\tau =$ constant, this represents a wave travelling with constant speed $1-k'$: hence the use of the word 'convection'. This analysis suggests that by changing to the moving frame of reference $\xi = x+(k'-1)\tau$, the $(k'-1)\partial c/\partial x$ term could be eliminated from the problem. This is indeed the case; the equation becomes

$$\frac{\partial c}{\partial \tau} = \frac{\partial^2 c}{\partial \xi^2} - kc - f\left(\xi - (k'-1)\tau\right),$$

in the variables ξ and τ.

The second term in (7.19) new to the diffusion equation, $-kc$, represents reaction (or discounting) at a rate proportional to c. If we balance this term against the $\partial c/\partial \tau$ we find that this balance gives $c = c_0 e^{-kt}$. This suggests this term could be eliminated by writing $c(x,\tau) = e^{-k\tau} w(x,\tau)$, to account for the exponential decay implied by it. This is the case, and the problem in the new dependent variable w is

$$\frac{\partial w}{\partial \tau} = \frac{\partial^2 w}{\partial \xi^2} - e^{k\tau} f\left(\xi - (k'-1)\tau\right).$$

Finally, and most importantly, the third new term in (7.19), $f(x)$, represents a consumption term if $f(x) < 0$ and a replenishment term if $f(x) > 0$. We may see this by balancing this term against the $\partial c/\partial \tau$ term. If $f(x) < 0$, then $\partial c/\partial \tau < 0$ in this balance and $c(x,\tau)$ decreases with τ, while if $f(x) > 0$, $c(x,\tau)$ increases.

7.7.2 A Local Analysis of the Free Boundary

We now ask how the free boundary $x = x_f(\tau)$ initially moves away from $x_f(0) = x_0$. We cannot solve the problem for the free boundary exactly, but we can find an asymptotic solution which is valid close to expiry.

In order to perform this analysis, which is local in both time and asset price, we look at equation (7.19) only near $x = x_0$, and for small values of τ. We approximate $f(x)$ by a Taylor series about x_0 (which corresponds to the final position of the optimal exercise boundary), namely

$$\begin{aligned} f(x) &= f(x_0) + f'(x_0)(x-x_0) + O((x-x_0)^2) \\ &\sim (x-x_0)f'(x_0) \\ &= -k(x-x_0). \end{aligned}$$

We need only keep the highest spatial derivative of c, i.e. $\partial^2 c/\partial x^2$, since this will be larger than c or $\partial c/\partial x$ in a region where c is changing rapidly. The upshot is an approximate local problem for c. Call this local solution

$\hat{c}(x,\tau)$; it satisfies

$$\frac{\partial \hat{c}}{\partial \tau} = \frac{\partial^2 \hat{c}}{\partial x^2} - k(x - x_0)$$

with

$$\hat{c} = \frac{\partial \hat{c}}{\partial x} = 0 \quad \text{on} \quad x = x_f(\tau), \quad x_f(0) = x_0.$$

It is fortunate that we can solve this local problem exactly. In fact, it has a similarity solution in terms of the variable

$$\xi = (x - x_0)/\sqrt{\tau},$$

of the form

$$\hat{c} = \tau^{3/2} c^*(\xi),$$

where c^* satisfies some yet to be determined ordinary differential equation. At the same time we try a free boundary of the form

$$x_f(\tau) = x_0 + \xi_0 \sqrt{\tau},$$

where ξ_0 is also yet to be determined. Although the free boundary is still unknown, we now have to find only the constant ξ_0 rather than a fully τ-dependent function $x_f(\tau)$.

Substituting the expression $\hat{c} = \tau^{3/2} c^*(\xi)$ into the partial differential equation for $\hat{c}$ gives

$$\sqrt{\tau}\left(\tfrac{3}{2}c^* - \tfrac{1}{2}\xi \frac{dc^*}{d\xi}\right) = \sqrt{\tau}\,\frac{d^2 c^*}{d\xi^2} - k(x - x_0),$$

and dividing through by $\sqrt{\tau}$ gives, as intended, an ordinary differential equation. We have

$$\frac{d^2 c^*}{d\xi^2} + \tfrac{1}{2}\xi \frac{dc^*}{d\xi} - \tfrac{3}{2}c^* = k\xi, \tag{7.23}$$

while the free boundary conditions

$$\hat{c} = \frac{\partial \hat{c}}{\partial x} = 0 \quad \text{on} \quad x = x_f(\tau)$$

reduce to

$$c^*(\xi_0) = \frac{dc^*}{d\xi}(\xi_0) = 0.$$

We also need to know how $\hat{c}(x,\tau)$ behaves for $\xi \to -\infty$. This follows from the behaviour of $\hat{c}(x,\tau)$ for large negative x: as $x \to -\infty$, $\partial^2 \hat{c}/\partial x^2 \to 0$ and so $\hat{c}(x,\tau)$ looks like $-kx\tau$. Thus

$$c^*(\xi) \sim -k\xi \quad \text{as} \quad \xi \to -\infty$$

(implying also that $\hat{c} = \tau^{\frac{3}{2}} c^* \sim \tau^{\frac{3}{2}}(x - x_0)/\tau^{\frac{1}{2}} \sim x\tau$ + smaller terms).

The first step in solving this two point boundary value problem for $c^*(\xi)$ is to find the general solution of the homogeneous ordinary differential equation

$$\frac{d^2 c^*}{d\xi^2} + \tfrac{1}{2}\xi \frac{dc^*}{d\xi} - \tfrac{3}{2}c^* = 0.$$

Fortunately one solution, $c_1^*(\xi)$ say, is easy to find. Trying simple low order polynomial solutions shows that

$$c_1^*(\xi) = \xi^3 + 6\xi$$

is an exact solution of the homogeneous equation. A second independent solution may be found by the method of reduction of order: we set $c_2^*(\xi) = c_1^*(\xi)a(\xi)$ and we find a *first* order ordinary differential equation for $a(\xi)$. The arithmetic is straightforward but tedious, and the result is that $c_2^*(\xi)$ is found to be

$$c_2^*(\xi) = (\xi^2 + 4)e^{-\frac{1}{4}\xi^2} + \tfrac{1}{2}(\xi^2 + 6\xi)\int_{-\infty}^{\xi} e^{-\frac{1}{4}s^2}\,ds.$$

Thus the general solution of the homogeneous equation is

$$c^*(\xi) = Ac_1^*(\xi) + Bc_2^*(\xi).$$

The second step in solving the original ordinary differential equation (7.23) is to observe that $c_p^*(\xi) = -k\xi$ is an exact solution. Thus the general solution is given by the sum of c_p^* and the general solution of the homogeneous equation, i.e.

$$c^*(\xi) = -k\xi + Ac_1^*(\xi) + Bc_2^*(\xi).$$

As $\xi \to -\infty$, $c_2^*(\xi) \to 0$, while $c_1^*(\xi)$ tends to ∞ like ξ^3. We know $c^*(\xi) \sim -k\xi$ as $\xi \to -\infty$ and therefore $A = 0$. Thus

$$c^*(\xi) = -k\xi + Bc_2^*(\xi).$$

The free boundary conditions $c^*(\xi_0) = 0$ and $dc^*(\xi_0)/d\xi = 0$ give us two equations for B and ξ_0. These are

$$Bc_2^*(\xi_0) = \xi_0, \quad \text{and} \quad B\frac{dc_2^*}{d\xi}(\xi_0) = 1.$$

Dividing these equations gives

$$\xi_0 \frac{dc_2^*}{d\xi}(\xi_0) = c_2^*(\xi_0)$$

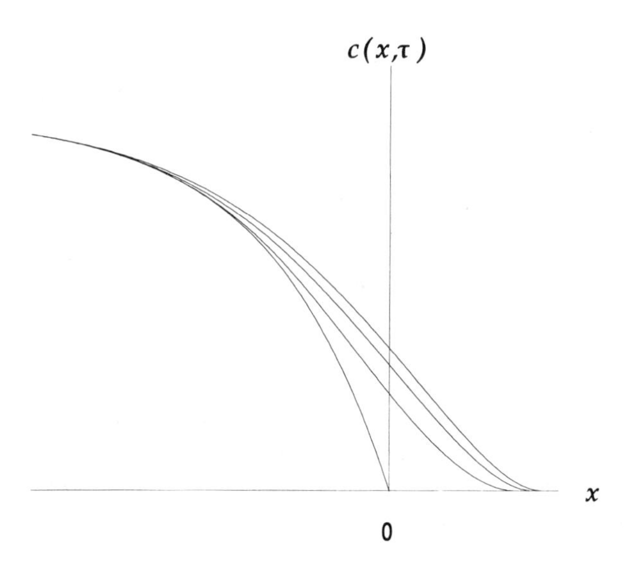

Figure 7.8. The function $c(x,\tau)$ at four different times, including expiry.

which, after some rearrangement, leads to the transcendental equation

$$\xi_0^3 e^{\frac{1}{4}\xi_0^2} \int_{-\infty}^{\xi_0} e^{-s^2/4} ds = 2(2 - \xi_0^2). \tag{7.24}$$

The constant B is then given by $B = \xi_0 / c_2^*(\xi_0)$.

Transcendental equations of the form (7.24) are characteristic features of similarity solutions of free boundary problems. It can be shown, for example by graphical methods, that this equation has just one root, and this can be found (by a numerical method such as the bisection algorithm or Newton iteration) to be

$$\xi_0 = 0.9034\ldots .$$

We have found a local solution $\hat{c}(x,\tau)$ which is a valid approximation to the American call problem for τ near to zero and x near to x_0.

Reverting to financial variables, we have shown that at expiry the optimal exercise price of an American call on a dividend-paying asset tends to the value rE/D_0. Also, from the local analysis, we know that as $t \to T$

$$S_f(t) \sim \frac{rE}{D}\left(1 + \xi_0\sqrt{\tfrac{1}{2}\sigma^2(T-t)} + \cdots\right)$$

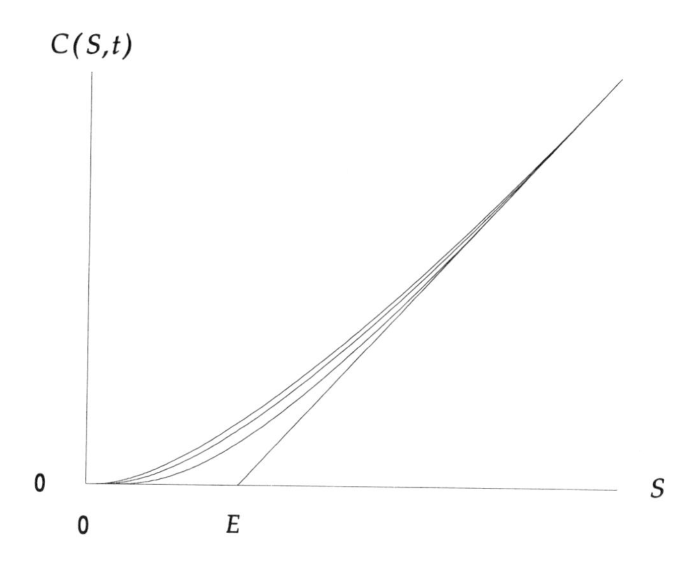

Figure 7.9. The option value $C(S,t)$ at the same four times as in Figure 7.8.

where ξ_0 is a 'universal constant' of call option pricing. Beyond this interesting fact, the local analysis is also important for the early stages of a numerical calculation where prices change rapidly: the effect of the rapid changes in $S_f(t)$ is felt throughout the solution region, not just near $S = S_f(T)$. In Figures 7.8 and 7.9 we show the values of $c(x,\tau)$ in dimensionless variables, and the original $C(S,t)$, prior to expiry.

Further Reading

- Readers who prefer a more formal derivation of the continuity of the delta at the optimal exercise price will find it in the books by Merton (1990) and Duffie (1992).
- For more on the theory of linear complementarity problems and variational inequalities see the books by Elliott & Ockendon (1982), Friedman (1988) and Kinderlehrer & Stampacchia (1980).
- Approximate solutions for American options have been found by Roll (1977), Whaley (1981), Barone-Adesi & Whaley (1987) and Johnson (1983).
- Hill & Dewynne (1990) and Crank (1984) discuss similarity solutions to free boundary problems for the diffusion equation.

Exercises

1. The **instalment option** has the same payoff as a vanilla call or put option; it may be European or American. Its unusual feature is that, as well as paying the initial premium, the holder must pay 'instalments' during the life of the option. The instalments may be paid either continuously or discretely. The holder can choose at any time to stop paying the instalments, at which point the contract is cancelled and the option ceases to exist. When instalments are paid continuously at a rate $L(t)$ per unit time, derive the differential equation satisfied by the option price. What new constraint must it satisfy? Formulate a free boundary problem for its value.

2. Consider American vanilla call and put options, with prices C and P. Derive the following inequalities (the second part of the last inequality is the version of put-call parity result appropriate for American options):

$$P \geq \max(E - S, 0), \qquad C \geq S - Ee^{-r(T-t)},$$

$$S - E \leq C - P \leq S - Ee^{-r(T-t)}.$$

Also show that, without dividends, it is never optimal to exercise an American call option.

3. Find the explicit solution to the obstacle problem (7.4) when the obstacle is $f(x) = \frac{1}{2} - x^2$. Repeat when $f(x) = \frac{1}{2} - \sin^2(\pi x/2)$; the free boundaries now have to be found numerically.

4. A space of functions $\mathcal{K}$ is said to be **convex** if, whenever $u \in \mathcal{K}$ and $v \in \mathcal{K}$, $(1-\lambda)u + \lambda v \in \mathcal{K}$ for all $0 \leq \lambda \leq 1$. Show that the space $\mathcal{K}$ of all piecewise continuously differentiable functions $v(x)$ $(-1 \leq x \leq 1)$ satisfying $v \geq f$ and $v(\pm 1) = 0$ is convex. (These functions are called **test functions.**)

The obstacle problem may also be formulated as: find the function u that minimises the energy

$$E[v] = \int_{-1}^{1} \tfrac{1}{2}(v')^2 \, dx$$

over all $v \in \mathcal{K}$. (This is the usual energy minimisation but with the constraint incorporated.) If u is the minimiser, and v is any test function, use the fact that $E[(1-\lambda)u + \lambda v] - E[u] \geq 0$ for all λ to show that

$$\int_{-1}^{1} u'(v-u)' \, dx \geq 0.$$

This is the **variational inequality** for the obstacle problem.

5. Set up the American call with dividends as a linear complementarity problem.

6. Transform the American cash-or-nothing call into a linear complementarity problem for the diffusion equation and show that the transformed payoff is

$$g(x,\tau) = be^{\frac{1}{2}(k+1)^2\tau + \frac{1}{2}(k-1)x}\mathcal{H}(x), \tag{E7.1}$$

where $b = B/E$.

Since in this case the free boundary is always at $x = 0$, the problem can be solved explicitly: do this. (Hint: put $u(x,\tau) = be^{\frac{1}{2}(k+1)^2\tau}X(x) + w(x,\tau)$ and choose $X(x)$ appropriately, then use images, as discussed in Chapter 12. Alternatively, use Laplace transforms or Duhamel's theorem.)

7. Set up the American call and put problems as linear complementarity problems using the original (S,t) variables. (This is not necessarily a good formulation as far as numerical solution is concerned.)

8. The function $u(x,\tau)$ satisfies the following free boundary problem with free boundary $x = x_f(\tau)$, where $x_f(0) = 0$:

$$\frac{\partial u}{\partial \tau} = \frac{\partial^2 u}{\partial x^2}, \qquad 0 < x < x_f(\tau),$$

$$u\big(x_f(\tau),\tau\big) = 0, \qquad \frac{\partial u}{\partial x}\,\big(x_f(\tau),\tau\big) = -\frac{dx_f}{d\tau},$$

$$u(0,\tau) = 1, \qquad t > 0.$$

Show that there is a similarity solution with $u(x,\tau) = u^*(x/\sqrt{\tau})$, $x_f(\tau) = \xi_0\sqrt{\tau}$, where ξ_0 satisfies the transcendental equation

$$\tfrac{1}{2}\xi_0 e^{\xi_0^2/4}\int_0^{\xi_0} e^{-s^2/4}\,ds = 1.$$

(This is a solution to the Stefan problem, in which $u(x,\tau)$ models the temperature in a pure material that melts at temperature $u = 0$; here a semi-infinite bar of the material is initially solid at the melting temperature, and melting is initiated by raising the temperature at $x = 0$ to 1 and holding it there, so that the region $0 < x < x_f(\tau)$ is liquid. The conditions at the free boundary express the facts that melting takes place at $u = 0$, and that the heat flux into the free boundary is balanced by the rate at which latent heat is taken up during the change of phase from solid to liquid.)

9. The function $c(x,\tau)$ satisfies the following problem in the region $0 < x < x_f(\tau)$, where $x_f(0) = 0$:

$$\frac{\partial c}{\partial \tau} = \frac{\partial^2 c}{\partial x^2} - 1, \qquad 0 < x < x_f(\tau),$$

$$c\big(x_f(\tau), \tau\big) = 0, \qquad \frac{\partial c}{\partial x}\big(x_f(\tau), \tau\big) = 0,$$

$$c(0,\tau) = \tau, \qquad \tau > 0.$$

Show that there is a similarity solution with $c(x,\tau) = \tau c^*(x/\sqrt{\tau})$, $x_f(\tau) = \xi_0\sqrt{\tau}$, where ξ_0 satisfies the transcendental equation

$$\tfrac{1}{2}\xi_0 e^{\xi_0^2/4} \int_0^{\xi_0} e^{-s^2/4}\, ds = 1.$$

10. Show that $u(x,\tau)$ and $c(x,\tau)$ of the previous two exercises are related by

$$u = \frac{\partial c}{\partial \tau}.$$

This requires you to show that the equations and boundary conditions all correspond. The only tricky part is to deal with the free boundary conditions for $c(x,\tau)$: differentiating the condition $c(x_f(\tau),\tau) = 0$ with respect to τ yields

$$\begin{aligned}\frac{\partial c}{\partial x}\big(x_f(\tau),\tau\big)\frac{dx_f}{d\tau} + \frac{\partial c}{\partial \tau} &= \frac{\partial c}{\partial x}\big(x_f(\tau),\tau\big)\frac{dx_f}{d\tau} + u\big(x_f(\tau),\tau\big) \\ &= 0 + u\big(x_f(\tau),\tau\big),\end{aligned}$$

which demonstrates that the condition $u\left(x_f(\tau),\tau\right) = 0$ holds; the second free boundary condition for u is obtained by differentiating $\partial c\big(x_f(\tau),\tau\big)/\partial x$ with respect to τ and using the partial differential equation for $c(x,\tau)$.

Part two

Numerical Methods

8 Finite-difference Methods

8.1 Introduction

Finite-difference methods are a means of obtaining numerical solutions to partial differential equations (as we see in this chapter) and linear complementarity problems (as we see in the following chapter). They constitute a very powerful and flexible technique and, if applied correctly, are capable of generating accurate numerical solutions to all of the models derived in this book, as well as to many other partial differential equations arising in both the physical and financial sciences. Needless to say, in such a brief introduction to the subject as we give here, we can only touch on the basics of finite differences; for more, see *Option Pricing*. Nevertheless, the underlying ideas generalise in a relatively straightforward manner to many more complicated problems.

As we saw in Chapter 5, once the Black–Scholes equation is reduced to the diffusion equation it is a relatively simple matter to find exact solutions (and then convert these back into financial variables). This is, of course, because the diffusion equation is a far simpler and less cluttered equation than the Black–Scholes equation. For precisely this reason also, it is a much simpler exercise to find numerical solutions of the diffusion equation and then, by a change of variables, to convert these into numerical solutions of the Black–Scholes equation, than it is to solve the Black–Scholes equation itself numerically. In this chapter, therefore, we concentrate on solving the diffusion equation using finite-differences. This allows us to introduce the fundamental ideas in as uncluttered an environment as possible.

This is not to say that one should not use finite differences to solve the Black–Scholes equation directly. There are many examples (particularly of multi-factor models) where it is not feasible or even not possible to reduce the problem to a constant coefficient diffusion equation (in one or more dimensions); in this case there is little choice but to use finite differences on the generalisations of the Black–Scholes equation. Direct application of finite-difference methods to the Black–Scholes equation is left to the exercises at the end of the chapter; the reader who has understood the underlying principles of finite differences should have no difficulty with these.

Recall from Sections 5.4 and 7.7.1 that, by using the change of variables (5.9) the Black–Scholes equation (3.9) for any European option can be transformed into the diffusion equation

$$\frac{\partial u}{\partial \tau} = \frac{\partial^2 u}{\partial x^2}.$$

The payoff function for the option determines the initial conditions for $u(x,\tau)$, and the boundary conditions for the option determine the conditions at infinity for $u(x,\tau)$ (that is, as $x \to \pm\infty$); for a put they are given by (3.13), (3.14) and (3.15); for a call they are given by (3.10), (3.11) and (3.12); and for a cash-or-nothing call the payoff is given on pages 38 and 82.

The values of the option $V(S,t)$, in financial variables, may be recovered from the non-dimensional $u(x,\tau)$ using (5.9), yielding

$$V = E^{\frac{1}{2}(1+k)} S^{\frac{1}{2}(1-k)} e^{\frac{1}{8}(k+1)^2\sigma^2(T-t)} u\left(\log\left(S/E\right), \tfrac{1}{2}\sigma^2(T-t)\right),$$

where $k = r/\frac{1}{2}\sigma^2$.

8.2 Finite-difference Approximations

The idea underlying finite-difference methods is to replace the partial derivatives occurring in partial differential equations by approximations based on Taylor series expansions of functions near the point or points of interest. For example, the partial derivative $\partial u/\partial\tau$ may be defined to be the limiting difference

$$\frac{\partial u}{\partial \tau}(x,\tau) = \lim_{\delta\tau\to 0} \frac{u(x,\tau+\delta\tau) - u(x,\tau)}{\delta\tau}.$$

If, instead of taking the limit $\delta\tau \to 0$, we regard $\delta\tau$ as nonzero but small, we obtain the approximation

$$\frac{\partial u}{\partial \tau}(x,\tau) \approx \frac{u(x,\tau+\delta\tau) - u(x,\tau)}{\delta\tau} + O(\delta\tau). \tag{8.1}$$

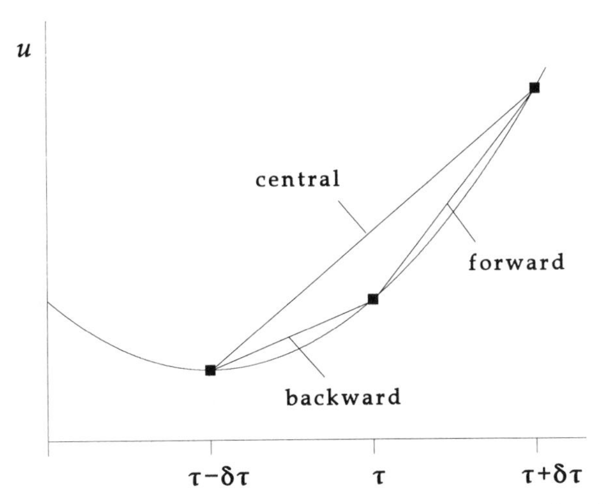

Figure 8.1 Forward-, backward- and central-difference approximations. The slopes of the lines are approximations to the tangent at (x, τ).

This is called a **finite-difference approximation** or a **finite difference** of $\partial u/\partial \tau$ because it involves small, but not infinitesimal, differences of the dependent variable u. This particular finite-difference approximation is called a **forward difference**, since the differencing is in the forward τ direction; only the values of u at τ and $\tau + \delta\tau$ are used. As the $O(\delta\tau)$ term suggests, the smaller $\delta\tau$ is, the more accurate the approximation.[1]

We also have

$$\frac{\partial u}{\partial \tau}(x, \tau) = \lim_{\delta\tau \to 0} \frac{u(x, \tau) - u(x, \tau - \delta\tau)}{\delta\tau},$$

so that the approximation

$$\frac{\partial u}{\partial \tau}(x, \tau) \approx \frac{u(x, \tau) - u(x, \tau - \delta\tau)}{\delta\tau} + O(\delta\tau) \tag{8.2}$$

is likewise a finite-difference approximation for $\partial u/\partial \tau$. We call this finite-difference approximation a **backward difference**.

We can also define **central differences** by noting that

$$\frac{\partial u}{\partial \tau}(x, \tau) = \lim_{\delta\tau \to 0} \frac{u(x, \tau + \delta\tau) - u(x, \tau - \delta\tau)}{2\,\delta\tau}.$$

[1] The $O(\delta\tau)$ term arises from a Taylor series expansion of $u(x, \tau + \delta\tau)$ about (x, τ); see Exercises 1–3.

This gives rise to the approximation

$$\frac{\partial u}{\partial \tau}(x,\tau) \approx \frac{u(x,\tau+\delta\tau) - u(x,\tau-\delta\tau)}{2\,\delta\tau} + O\Big((\delta\tau)^2\Big). \tag{8.3}$$

Figure 8.1 shows a geometric interpretation of these three types of finite differences. Note that central differences are more accurate (for small $\delta\tau$) than either forward or backward differences; this is also suggested by Figure 8.1. (For the analysis of the accuracy of finite-difference approximations, see Exercises 1–3 at the end of the chapter.)

When applied to the the diffusion equation, forward- and backward-difference approximations for $\partial u/\partial\tau$ lead to **explicit** and **fully implicit** finite-difference schemes, respectively. Central differences of the form (8.3) are never used in practice because they always lead to bad numerical schemes (specifically, schemes that are inherently unstable). Central differences of the form

$$\frac{\partial u}{\partial \tau} \approx \frac{u(x,\tau+\delta\tau/2) - u(x,\tau-\delta\tau/2)}{\delta\tau} + O\Big((\delta\tau)^2\Big) \tag{8.4}$$

arise in the **Crank–Nicolson** finite-difference scheme.

We can define finite-difference approximations for the x-partial derivative of u in exactly the same way. For example, the central finite-difference approximation is easily seen to be[2]

$$\frac{\partial u}{\partial x}(x,\tau) \approx \frac{u(x+\delta x,\tau) - u(x-\delta x,\tau)}{2\,\delta x} + O\Big((\delta x)^2\Big). \tag{8.5}$$

For second partial derivatives, such as $\partial^2 u/\partial x^2$, we can define a symmetric finite-difference approximation as the forward difference of backward-difference approximations to the first derivative or as the backward difference of forward-difference approximations to the first derivative. In either case we obtain the **symmetric central-difference** approximation

$$\frac{\partial^2 u}{\partial x^2}(x,\tau) \approx \frac{u(x+\delta x,\tau) - 2u(x,\tau) + u(x-\delta x,\tau)}{(\delta x)^2} + O\Big((\delta x)^2\Big). \tag{8.6}$$

Although there are other approximations, this approximation to $\partial^2 u/\partial x^2$ is preferred, as its symmetry preserves the reflectional symmetry of the second partial derivative; it is left invariant by reflections of the form $x \mapsto -x$. It is also more accurate than other similar approximations.

[2] Although central differences of the form (8.3) are never used for τ- or t-partial derivatives, differences of the form (8.5) for x- or S-partial derivatives are used; see, for example, Exercises 9 and 13.

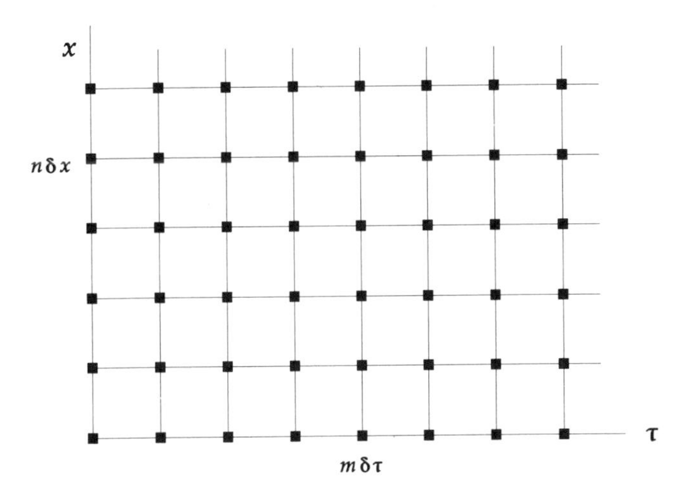

Figure 8.2 The mesh for a finite-difference approximation.

8.3 The Finite-difference Mesh

To continue with the finite-difference approximation to the diffusion equation we divide the x-axis into equally spaced **nodes** a distance δx apart, and the τ-axis into equally spaced nodes a distance $\delta\tau$ apart. This divides the (x,τ) plane into a mesh, where the **mesh points** have the form $(n\,\delta x, m\,\delta\tau)$; see Figure 8.2. We then concern ourselves only with the values of $u(x,\tau)$ at mesh points $(n\,\delta x, m\,\delta\tau)$; see Figure 8.3. We write

$$u_n^m = u(n\,\delta x, m\,\delta\tau) \tag{8.7}$$

for the value of $u(x,\tau)$ at the mesh point $(n\,\delta x, m\,\delta\tau)$.

8.4 The Explicit Finite-difference Method

Consider the general form of the transformed Black–Scholes model for the value of a European option,

$$\frac{\partial u}{\partial \tau} = \frac{\partial^2 u}{\partial x^2},$$

with boundary and initial conditions

$$\begin{gathered} u(x,\tau) \sim u_{-\infty}(x,\tau), \quad u(x,\tau) \sim u_{\infty}(x,\tau) \quad \text{as} \quad x \to \pm\infty \\ u(x,0) = u_0(x). \end{gathered} \tag{8.8}$$

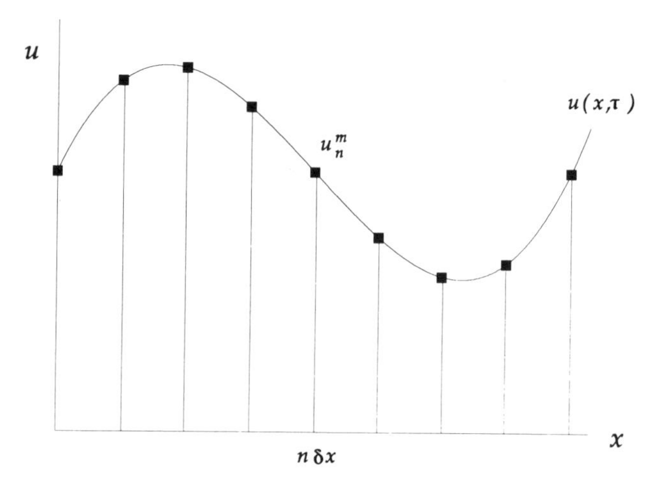

Figure 8.3 The finite-difference approximation at a fixed time-step.

We use the notation $u_{-\infty}(\tau)$, $u_{\infty}(\tau)$ and $u_0(x)$ to emphasise that the following does not in any way depend on the particular boundary and initial conditions involved. (For puts, calls and cash-or-nothing calls these are given as above.)

Confining our attention to values of u at mesh points, and using a forward difference for $\partial u/\partial \tau$, equation (8.1), and a symmetric central difference for $\partial^2 u/\partial x^2$, equation (8.6), we find that the diffusion equation becomes

$$\frac{u_n^{m+1} - u_n^m}{\delta \tau} + O(\delta \tau) = \frac{u_{n+1}^m - 2u_n^m + u_{n-1}^m}{(\delta x)^2} + O((\delta x)^2). \tag{8.9}$$

Ignoring terms of $O(\delta\tau)$ and $O\big((\delta x)^2\big)$, we can rearrange this to give the difference equations

$$u_n^{m+1} = \alpha u_{n+1}^m + (1 - 2\alpha) u_n^m + \alpha u_{n-1}^m \tag{8.10}$$

where

$$\alpha = \frac{\delta \tau}{(\delta x)^2}. \tag{8.11}$$

(Note that, whereas (8.9) is exact, albeit vague about the error terms, (8.10) is only approximate.)

If, at time-step m, we know u_n^m for all values of n we can explicitly calculate u_n^{m+1}. This is why this method is called explicit. Note that u_n^{m+1} depends only on u_{n+1}^m, u_n^m and u_{n-1}^m, as shown in Figure 8.4. This

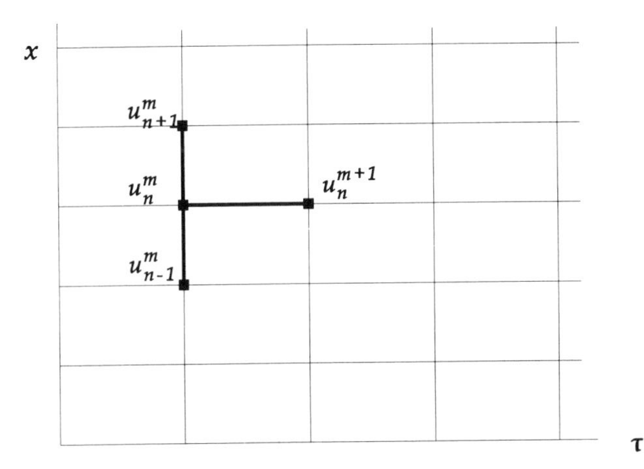

Figure 8.4 Explicit finite-difference discretisation.

figure also suggests that (8.10) may be considered as a random walk on a regular lattice, where u_n^m denotes the probability of a marker being at position n at time-step m, and α denotes the probability of it moving to the right or left by one unit and $(1-2\alpha)$ is the probability of it staying put.

If we choose a constant x-spacing δx, we cannot solve the problem for all $-\infty < x < \infty$ without taking an infinite number of x-steps. We get around this problem by taking a finite, but suitably large, number of x-steps. We restrict our attention to the interval

$$N^-\delta x \le x \le N^+\delta x$$

where $-N^-$ and N^+ are large positive integers.

To obtain the finite-difference solution for the option price, we divide the non-dimensional time to expiry of the option, $\frac{1}{2}\sigma^2 T$, into M equal time-steps so that

$$\delta\tau = \tfrac{1}{2}\sigma^2 T/M.$$

We then solve the difference equations (8.10) for $N^- < n < N^+$ and $0 < m \le M$, and use the boundary conditions from (8.8) to determine $u_{N^+}^m$ and $u_{N^-}^m$:

$$\begin{aligned} u_{N^-}^m &= u_{-\infty}(N^-\delta x, m\,\delta\tau), \quad 0 < m \le M, \\ u_{N^+}^m &= u_{\infty}(N^+\delta x, m\,\delta\tau), \quad 0 < m \le M. \end{aligned} \tag{8.12}$$

```
explicit_fd( values,dx,dt,M,Nplus,Nminus )
{
  a = dt/(dx*dx);

  for( n=Nminus; n<=Nplus; ++n )
    oldu[n] = pay_off( n*dx );

  for( m=1; m<=M; ++m )
  {
    tau = m*dt;

    newu[Nminus] = u_m_inf( Nminus*dx,tau );
    newu[ Nplus] = u_p_inf(  Nplus*dx,tau );

    for( n=Nminus+1; n<Nplus; ++n )
      newu[n] = oldu[n]
                + a*(oldu[n-1]-2*oldu[n]+oldu[n+1]);

    for( n=Nminus; n<=Nplus; ++n )
      oldu[n] = newu[n];
  }

  for( n=Nminus; n<=Nplus; ++n )
    values[n] = oldu[n];
}
```

Figure 8.5 Pseudo-code for the explicit finite-difference solution of the diffusion equation. The variables are **a** $= \alpha$, **tau** $= \tau$, **Nminus** $= N^-$, **Nplus** $= N^+$. The values of u_n^m are stored in the array **oldu**[], and the values of u_n^{m+1} are stored in the array **newu**[]. Initially the values of u_n^0 are put in **oldu**[]. Once all of the u_n^{m+1} are found they are copied back into **oldu**[] and the process is repeated until all the required time-steps have been completed. The numerical solution is copied into **values**[] and returned to the calling program.

To start the iterative procedure we use the initial condition from (8.8):

$$u_n^0 = u_0(n\,\delta x), \quad N^- \le n \le N^+. \tag{8.13}$$

As the equations determining u_n^{m+1} in terms of the u_n^m are explicit, this process can be easily coded for a computer to solve; a pseudo-code is given in Figure 8.5.

In Figure 8.6 we compare explicit finite-difference solutions for a European put with the exact Black–Scholes formula (note that we have transformed back into financial variables using (5.9)). We have deliberately chosen to regard α and $\delta\tau$ as variable, rather than the more obvious choice of δx and $\delta\tau$, to illustrate an extremely important point. When $\alpha = 0.25$ and $\alpha = 0.5$ there is good agreement between computed

S	$\alpha = 0.25$	$\alpha = 0.50$	$\alpha = 0.52$	exact
0.00	9.7531	9.7531	9.7531	9.7531
2.00	7.7531	7.7531	7.7531	7.7531
4.00	5.7531	5.7531	5.7531	5.7531
6.00	3.7532	3.7532	2.9498	3.7532
7.00	2.7567	2.7567	−17.4192	2.7568
8.00	1.7986	1.7985	95.3210	1.7987
9.00	0.9879	0.9879	350.5603	0.9880
10.00	0.4418	0.4419	625.0347	0.4420
11.00	0.1605	0.1607	−457.3122	0.1606
12.00	0.0483	0.0483	−208.9135	0.0483
13.00	0.0124	0.0123	40.5813	0.0124
14.00	0.0028	0.0027	−15.2150	0.0028
15.00	0.0006	0.0005	−3.1582	0.0006
16.00	0.0001	0.0001	0.7365	0.0001

Figure 8.6 Comparison of exact Black–Scholes solution and explicit finite-difference solutions for a European put with $E = 10$, $r = 0.05$, $\sigma = 0.20$ and with six months to expiry. Note the effect of taking $\alpha > \frac{1}{2}$.

and exact solutions, whereas when $\alpha = 0.52$, the computed solution is nonsensical. This illustrates the **stability problem** for explicit finite differences.

The stability problem arises because we are using *finite precision computer arithmetic* to solve the difference equations (8.10). This introduces rounding errors into the *numerical* solution of (8.10). The system (8.10) is said to be **stable** if these rounding errors are not magnified at each iteration. If the rounding errors do grow in magnitude at each iteration of the solution procedure, then (8.10) is said to be **unstable**.

It can be shown (see Exercise 5 at the end of the chapter) that the system (8.10) is:

- stable if $0 < \alpha \leq \frac{1}{2}$ (**stability condition**);
- unstable if $\alpha > \frac{1}{2}$ (**instability condition**).

If we regard the explicit finite-difference equations in terms of a random walk, instability corresponds to the presence of negative probabilities (specifically, the probability of a marker staying put, $1 - 2\alpha$, is negative).

The stability condition puts severe constraints on the size of time-steps. For stability we must have

$$0 < \frac{\delta\tau}{(\delta x)^2} \le \tfrac{1}{2}.$$

Thus if we start with a stable solution on a mesh and double the number of x-mesh points, for example to improve accuracy, we must quarter the size of the time-step. Each time-step then takes twice as long (twice as many x-mesh spacings) and there are four times as many time-steps. Thus, doubling the number of x-mesh points means that finding the solution takes eight times as long.

It can be shown that the numerical solution of the finite-difference equations converges to the exact solution of the diffusion equation as $\delta x \to 0$ and $\delta\tau \to 0$, in the sense that

$$u_n^m \to u(n\,\delta x, m\,\delta\tau),$$

if and only if the explicit finite-difference method is stable; the proof of this is beyond the scope of this text and we refer the reader to *Option Pricing*.

Note that because of the method's independence from the initial and boundary conditions, the explicit finite-difference method is easily adapted to deal with more general binary and barrier options (see Exercise 7).

8.5 Implicit Finite-difference Methods

Implicit finite-difference methods are used to overcome the stability limitations imposed by the restriction $0 < \alpha \le \frac{1}{2}$, which applies to the explicit method. Implicit methods allow us to use a large number of x-mesh points without having to take ridiculously small time-steps.

Implicit methods require the solution of *systems* of equations. We consider the techniques of LU decomposition and SOR for solving numerically these systems. By using these techniques implicit methods may be made almost as efficient as the explicit method in terms of arithmetical operations per time-step.[3] As fewer time-steps need to be taken, implicit finite-difference methods are usually more efficient overall than explicit methods. We shall consider both the fully implicit and Crank–Nicolson methods.

[3] In their most efficient forms explicit and implicit methods require $O(2N)$ and $O(4N)$ arithmetical operations per time-step, respectively, where N is the number of x-grid points.

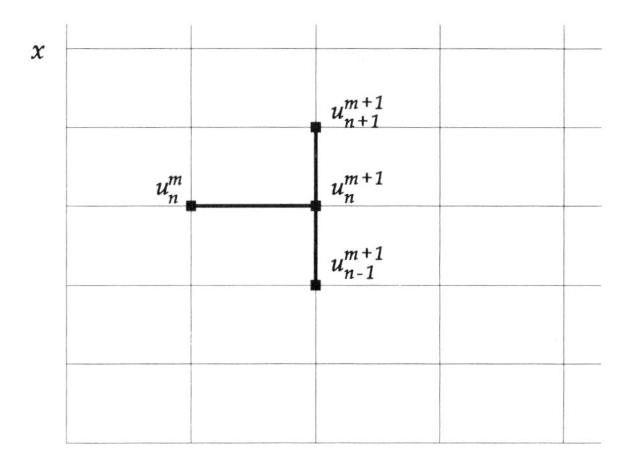

Figure 8.7 Implicit finite-difference discretisation.

8.6 The Fully-implicit Method

The fully-implicit finite-difference scheme, which is usually known as the **implicit finite-difference** method, uses the backward-difference approximation (8.2) for the $\partial u/\partial\tau$ term and the symmetric central-difference approximation (8.6) for the $\partial^2 u/\partial x^2$ term. This leads to the equation

$$\frac{u_n^{m+1} - u_n^m}{\delta\tau} + O(\delta\tau) = \frac{u_{n+1}^{m+1} - 2u_n^{m+1} + u_{n-1}^{m+1}}{(\delta x)^2} + O\left((\delta x)^2\right),$$

where we are using the same notation as in the previous sections. Again, we neglect terms of $O(\delta\tau)$, $O((\delta x)^2)$ and higher and, after some rearrangement, we find the implicit finite-difference equations

$$-\alpha u_{n-1}^{m+1} + (1+2\alpha)u_n^{m+1} - \alpha u_{n+1}^{m+1} = u_n^m. \tag{8.14}$$

As before, the space-step and the time-step are related through the parameter α, defined by (8.11). In the implicit finite-difference equation (8.14), u_n^{m+1}, u_{n-1}^{m+1} and u_{n+1}^{m+1} all depend on u_n^m in an *implicit* manner; the new values cannot immediately be separated out and solved for explicitly in terms of the old values. The scheme is illustrated in Figure 8.7.

Let us consider the European option problem discussed in the previous sections. We assume that we can truncate the infinite mesh at $x = N^-\delta x$ and $x = N^+\delta x$, and take N^- and N^+ sufficiently large so that no significant errors are introduced. As before we find the u_n^0 using (8.13) and $u_{N^-}^{m+1}$ and $u_{N^+}^{m+1}$ using (8.12). The problem is then to find the u_n^{m+1} for $m \geq 0$ and $N^- < n < N^+$ from (8.14).

We can write (8.14) as the linear system

$$\begin{pmatrix} 1+2\alpha & -\alpha & 0 & \cdots & 0 \\ -\alpha & 1+2\alpha & -\alpha & & 0 \\ 0 & -\alpha & \ddots & \ddots & \\ \vdots & & \ddots & \ddots & -\alpha \\ 0 & 0 & & -\alpha & 1+2\alpha \end{pmatrix} \begin{pmatrix} u_{N^-+1}^{m+1} \\ \vdots \\ u_0^{m+1} \\ \vdots \\ u_{N^+-1}^{m+1} \end{pmatrix}$$
$$= \begin{pmatrix} u_{N^-+1}^{m} \\ \vdots \\ u_0^{m} \\ \vdots \\ u_{N^+-1}^{m} \end{pmatrix} + \alpha \begin{pmatrix} u_{N^-}^{m+1} \\ 0 \\ \vdots \\ 0 \\ u_{N^+}^{m+1} \end{pmatrix} = \begin{pmatrix} b_{N^-+1}^{m} \\ \vdots \\ b_0^{m} \\ \vdots \\ b_{N^+-1}^{m} \end{pmatrix}, \quad \text{say.} \tag{8.15}$$

The vector on the right-hand side of the middle term in this equation arises from the end equations, for example

$$(1+2\alpha)u_{N^+-1}^{m+1} - \alpha u_{N^+-2}^{m+1} = u_{N^+-1}^{m} + \alpha u_{N^+}^{m+1}.$$

We can write (8.15) in the more compact form

$$\mathbf{M}\boldsymbol{u}^{m+1} = \boldsymbol{b}^m \tag{8.16}$$

where $\boldsymbol{u}^{m+1}$ and $\boldsymbol{b}^m$ denote the $(N^+ - N^- - 1)$-dimensional vectors

$$\boldsymbol{u}^{m+1} = (u_{N^-+1}^{m+1}, \ldots, u_{N^+-1}^{m+1}), \quad \boldsymbol{b}^m = \boldsymbol{u}^m + \alpha(u_{N^-}^{m+1}, 0, 0, \ldots, 0, u_{N^+}^{m+1}),$$

and $\mathbf{M}$ is the $(N^+ - N^- - 1)$-square symmetric matrix given in (8.15).

It can be shown that, for $\alpha \geq 0$, $\mathbf{M}$ is invertible and so in principle

$$\boldsymbol{u}^{m+1} = \mathbf{M}^{-1}\boldsymbol{b}^m, \tag{8.17}$$

where $\mathbf{M}^{-1}$ is the inverse of $\mathbf{M}$. We can therefore find $\boldsymbol{u}^{m+1}$ given $\boldsymbol{b}^m$, which in turn may be found from $\boldsymbol{u}^m$ and the boundary conditions. As the initial condition determines $\boldsymbol{u}^0$, we can find each $\boldsymbol{u}^{m+1}$ sequentially.

8.6.1 Practical Considerations

In practice there are far more efficient solution techniques than matrix inversion. The matrix $\mathbf{M}$ has the property that it is **tridiagonal**; that is, only the diagonal, super-diagonal and sub-diagonal elements are non-zero. This has a number of important consequences.

First, it means that we do not have to store all the zeros, just the non-zero elements. The inverse of $\mathbf{M}$, $\mathbf{M}^{-1}$, is not tridiagonal and requires a great deal more storage.[4]

Second, the tridiagonal structure of $\mathbf{M}$ means that there are highly efficient algorithms for solving (8.16) in $O(N)$ arithmetic operations per solution (specifically, about $4N$ operations). We now discuss two of these algorithms, **LU decomposition** and **SOR**.

8.6.2 The LU Method

In LU decomposition we look for a decomposition of the matrix $\mathbf{M}$ into a product of a lower triangular matrix $\mathbf{L}$ and an upper triangular matrix $\mathbf{U}$, namely $\mathbf{M} = \mathbf{LU}$, of the form

$$\begin{pmatrix} 1+2\alpha & -\alpha & 0 & \cdots & 0 \\ -\alpha & 1+2\alpha & -\alpha & & \vdots \\ 0 & -\alpha & \ddots & \ddots & 0 \\ \vdots & & \ddots & \ddots & -\alpha \\ 0 & \cdots & 0 & -\alpha & 1+2\alpha \end{pmatrix} =$$

$$\begin{pmatrix} 1 & 0 & 0 & \cdots & 0 \\ \ell_{N^-+1} & 1 & \ddots & & \vdots \\ 0 & \ddots & \ddots & \ddots & 0 \\ \vdots & & \ddots & \ddots & 0 \\ 0 & \cdots & 0 & \ell_{N^+-2} & 1 \end{pmatrix} \begin{pmatrix} y_{N^-+1} & z_{N^-+1} & 0 & \cdots & 0 \\ 0 & y_{N^-+2} & \ddots & & \vdots \\ 0 & \ddots & \ddots & \ddots & 0 \\ \vdots & & \ddots & \ddots & z_{N^+-2} \\ 0 & \cdots & 0 & 0 & y_{N^+-1} \end{pmatrix}. \tag{8.18}$$

In order to determine the quantities ℓ_n, y_n and z_n (and observe that these have only to be calculated once) we simply multiply together the two matrices on the right-hand side of (8.18) and equate the result to the left-hand side. After some simple manipulation we find that

$$\begin{aligned} y_{N^-+1} &= (1+2\alpha), \\ y_n &= (1+2\alpha) - \alpha^2/y_{n-1}, \quad n = N^- + 2, \ldots, N^+ - 1, \\ z_n &= -\alpha, \quad \ell_n = -\alpha/y_n, \quad n = N^- + 1, \ldots, N^+ - 2. \end{aligned} \tag{8.19}$$

[4] If N is the dimension of the system, storing $\mathbf{M}^{-1}$ requires N^2 real numbers, whereas storing the non-zero elements of $\mathbf{M}$ requires $3N - 2$. Furthermore, the most efficient means of inverting $\mathbf{M}$ requires $O(N^2)$ operations, and the matrix multiplication required to find $\mathbf{M}^{-1}\boldsymbol{b}^m$ requires a further $O(N^2)$ operations.

This also shows that the only quantities we need to calculate and save are the y_n, $n = N^- + 1, \dots, N^+ - 1$.

The original problem $\mathbf{M}\boldsymbol{u}^{m+1} = \boldsymbol{b}^m$ can be written as $\mathbf{L}(\mathbf{U}\boldsymbol{u}^{m+1}) = \boldsymbol{b}^m$, which may then be broken down into two simpler subproblems,

$$\mathbf{L}\boldsymbol{q}^m = \boldsymbol{b}^m, \quad \mathbf{U}\boldsymbol{u}^{m+1} = \boldsymbol{q}^m,$$

where $\boldsymbol{q}^m$ is an intermediate vector. Thus, having eliminated the ℓ_n from the lower triangular matrix and the z_n from the upper triangular matrix using (8.19), the solution procedure is to solve the two subproblems

$$\begin{pmatrix} 1 & 0 & 0 & \cdots & 0 \\ -\dfrac{\alpha}{y_{N^-+1}} & 1 & 0 & & \vdots \\ 0 & -\dfrac{\alpha}{y_{N^-+2}} & \ddots & \ddots & 0 \\ \vdots & & \ddots & \ddots & 0 \\ 0 & \cdots & 0 & -\dfrac{\alpha}{y_{N^+-2}} & 1 \end{pmatrix} \begin{pmatrix} q^m_{N^-+1} \\ q^m_{N^-+2} \\ \vdots \\ q^m_{N^+-2} \\ q^m_{N^+-1} \end{pmatrix} = \begin{pmatrix} b^m_{N^-+1} \\ b^m_{N^-+2} \\ \vdots \\ b^m_{N^+-2} \\ b^m_{N^+-1} \end{pmatrix} \tag{8.20}$$

and

$$\begin{pmatrix} y_{N^-+1} & -\alpha & 0 & \cdots & 0 \\ 0 & y_{N^-+2} & -\alpha & & \vdots \\ 0 & 0 & \ddots & \ddots & 0 \\ \vdots & & \ddots & y_{N^+-2} & -\alpha \\ 0 & \cdots & 0 & 0 & y_{N^+-1} \end{pmatrix} \begin{pmatrix} u^{m+1}_{N^-+1} \\ u^{m+1}_{N^-+2} \\ \vdots \\ u^{m+1}_{N^+-2} \\ u^{m+1}_{N^+-1} \end{pmatrix} = \begin{pmatrix} q^m_{N^-+1} \\ q^m_{N^-+2} \\ \vdots \\ q^m_{N^+-2} \\ q^m_{N^+-1} \end{pmatrix}. \tag{8.21}$$

The intermediate quantities q^m_n are easily found by forward substitution. We can read off the value of $q^m_{N^-+1}$ directly, while any other equation in the system relates only q^m_n and q^m_{n-1}. If we solve the system in increasing n-indicial order, we have q^m_{n-1} available at the time we have to solve for q^m_n. Thus we can find q^m_n easily:

$$q^m_{N^-+1} = b^m_{N^-+1}, \quad q^m_n = b^m_n + \frac{\alpha q^m_{n-1}}{y_{n-1}}, \quad n = N^-+2, \dots, N^+-1. \tag{8.22}$$

Similarly, solving (8.21) for the u^m_n (given that we have found the intermediate q^m_n) is easily achieved by backward substitution. This time it is $u^{m+1}_{N^+-1}$ that can be read off directly, and if we solve in decreasing

```
lu_find_y( y,a,Nminus,Nplus )
{
  asq = a*a;
  y[Nminus+1] = 1+2*a;

  for( n=Nminus+2; n<Nplus; ++n )
  {
    y[n] = 1+2*a - asq/y[n-1];
    if (y[n]==0) return(SINGULAR);
  }
  return( OK );
}

lu_solver( u,b,y,a,Nminus,Nplus )
{
/* Must call lu_find_y before using this */

  q[Nminus+1] = b[Nminus+1];

  for( n=Nminus+2; n<Nplus; ++n )
    q[n] = b[n]+a*q[n-1]/y[n-1];

  u[Nplus-1] = q[Nplus-1]/y[Nplus-1];

  for( n=Nplus-2; n>Nminus; --n )
    u[n] = (q[n]+a*u[n+1])/y[n];
}
```

Figure 8.8 Pseudo-code for LU tridiagonal solver. Variables are `a` $= \alpha$, `asq` $=\alpha^2$, `Nplus` $= N^+$, `Nminus` $= N^-$, `b[n]` $= b_n^m$, `q[n]` $= q_n^m$, `u[n]` $= u_n^{m+1}$, `y[n]` $= y_n$. The routine solves the problem only for `Nminus` $+$ `1` $\leq$ `n` $\leq$ `Nplus` $-$ `1`; the values at `Nplus` and `Nminus` are assumed to have been set by the calling routine. The calling routine must call `lu_find_y` before it calls `lu_solver` in order to set up `y[]`.

n-indicial order we can find all of the u_n^m in an equally simple manner:

$$u_{N^+-1}^{m+1} = \frac{q_{N^+-1}^m}{y_{N^+-1}}, \quad u_n^{m+1} = \frac{q_n^m + \alpha u_{n+1}^{m+1}}{y_n}, \quad n = N^+ - 2, \ldots, N^- + 1. \tag{8.23}$$

Equations (8.19), (8.22) and (8.23) define the LU algorithm:

- find the y_n using (8.19); [5]
- given the vector $\boldsymbol{b}^m$, use (8.22) to find the vector $\boldsymbol{q}^m$;
- use (8.23) to find the required solution $\boldsymbol{u}^{m+1}$.

[5] Note that these depend only on the matrix $\mathbf{M}$ and not on $\boldsymbol{b}^m$. We propose solving the system $\mathbf{M}\boldsymbol{u}^{m+1} = \boldsymbol{b}^m$ for many time-steps. We have, however, to find the y_n once only; they are the same for each time-step.

The algorithm is illustrated by the pseudo-code in Figure 8.8. (Note that the ideas above can also be applied to quite general tridiagonal systems where the super-, sub- and diagonal elements vary with position in the matrix. Such matrices arise if the implicit method is used directly on the Black–Scholes equation; see Exercise 14.)

8.6.3 The SOR Method

The LU method discussed in the previous section is a *direct* method for solving the system (8.16) in that it aims to find the unknowns exactly and in one pass. An alternative strategy is to employ an *iterative* method. Iterative methods differ from direct methods in that one starts with a guess for the solution and successively improves it until it converges to the exact solution (or is near enough to the exact solution). In a direct method, one obtains the solution without any iteration. An advantage of iterative methods over direct methods is that they generalise in straightforward ways to American option problems and nonlinear models involving transaction costs, whereas direct methods do not. Another advantage is that they are easier to program. On the other hand, a disadvantage of iterative methods is that, in the context of European option problems, they are somewhat slower than direct methods.[6]

The acronym **SOR** stands for Successive Over-Relaxation, and the SOR algorithm is an example of an iterative method. It is a refinement of another iterative method known as the **Gauss–Seidel** method, which in turn is a development of the **Jacobi** method. It is easiest to explain the SOR method by first describing these other two related but simpler methods. All three iterative methods rely on the fact that the system (8.14) (or 8.15 or (8.16)) may be written in the form

$$u_n^{m+1} = \frac{1}{1+2\alpha}\Big(b_n^m + \alpha(u_{n-1}^{m+1} + u_{n+1}^{m+1})\Big). \qquad (8.24)$$

This equation is simply a rearrangement of either of (8.14), (8.15) or (8.16) that isolates the diagonal terms in the problem on the left-hand side of the equation.

The idea behind the Jacobi method is to take some initial guess for u_n^{m+1} for $N^- + 1 \le n \le N^+ - 1$ (a good initial guess is the values of

[6] The LU algorithm described in the previous section requires about $4N$ operations per time-step. The SOR algorithm described in this section requires $4N \times$ (number of iterations) per time-step. Typically the number of iterations is of the order of two or three.

u from the previous step, i.e. u_n^m) and substitute these into the right-hand side of (8.24) to obtain a new guess for u_n^{m+1} (from the left-hand side). The process is then repeated until the approximations cease to change (or cease to change significantly). When this happens we have the solution.

Formally we can define the **Jacobi** method as follows. Let $u_n^{m+1,k}$ be the k-th iterate for u_n^{m+1}. Thus, the initial guess is denoted by $u_n^{m+1,0}$ and as $k \to \infty$ we expect $u_n^{m+1,k} \to u_n^{m+1}$. Then, given $u_n^{m+1,k}$, we calculate $u_n^{m+1,k+1}$ using a modified form of (8.24), namely,

$$u_n^{m+1,k+1} = \frac{1}{1+2\alpha}\left(b_n^m + \alpha(u_{n-1}^{m+1,k} + u_{n+1}^{m+1,k})\right), \quad N^- < n < N^+. \tag{8.25}$$

The whole process is repeated until a measure of the error such as

$$\|\boldsymbol{u}^{m+1,k+1} - \boldsymbol{u}^{m+1,k}\|^2 = \sum_n \left(u_n^{m+1,k+1} - u_n^{m+1,k}\right)^2$$

becomes sufficiently small for us to regard any further iterations as unnecessary; then we take $u_n^{m+1,k+1}$ as the value of u_n^{m+1}. The method is known to converge to the correct solution for any value of $\alpha > 0$, although a full discussion of the convergence of the algorithm is beyond the scope of this text.

The **Gauss–Seidel** method is a development of the Jacobi method. It relies on the fact that when we come to calculate $u_n^{m+1,k+1}$ in (8.25) we have already found $u_{n-1}^{m+1,k+1}$. In the Gauss–Seidel method we use this value instead of $u_{n-1}^{m+1,k}$. Thus, the Gauss–Seidel method is identical to the Jacobi method, except that (8.25) is replaced by

$$u_n^{m+1,k+1} = \frac{1}{1+2\alpha}\left(b_n^m + \alpha(u_{n-1}^{m+1,k+1} + u_{n+1}^{m+1,k})\right), \quad N^- < n < N^+. \tag{8.26}$$

The difference may be summarised by saying that, in the Gauss–Seidel method, we use an updated guess immediately when it becomes available, while in the Jacobi method we use updated guesses only when they are *all* available. One practical consequence of using the most recent information (i.e. $u_n^{m+1,k+1}$ rather than $u_n^{m+1,k}$) is that the Gauss–Seidel method converges more rapidly than the Jacobi method and is therefore more efficient. In fact, the Gauss–Seidel method is even more efficient, as the updating is achieved by overwriting old iterates, whereas in the Jacobi method the old and new iterates have to be stored separately until all the new iterates are found (and then copied over the old iterates).

```
SOR_solver( u,b,Nminus,Nplus,a,omega,eps,loops )
{
  loops = 0;
  do
  {
    error = 0.0;
    for( n=Nminus+1; n<Nplus; ++n )
    {
      y = ( b[n]+a*(u[n-1]+u[n+1]) )/(1+2*a);
      y = u[n]+omega*(y-u[n]);
      error += (u[n]-y)*(u[n]-y);
      u[n]=y;
    }
    ++loops;
  }
  while ( error > eps );
  return(loops);
}
```

Figure 8.9 Pseudo-code for SOR algorithm for a European option problem. Here `a` $= \alpha$, `Nplus` $= N^+$, `Nminus` $= N^-$, `b[n]` $= b_n^m$, `u[n]` $= u_n^{m+1,k}$ or $u_n^{m+1,k+1}$, `y` $= y_n^{m+1,k+1}$, `omega` $= \omega$ and `eps` is the desired error tolerance. The routine over-writes old iterates as soon as a new iterate is generated; thus for a given value of `n` in the loop, `u[n − 1]`, `u[n − 2]`, and so forth, will contain $u_{n-1}^{m+1,k+1}$, $u_{n-2}^{m+1,k+1}$, etc., whereas `u[n + 1]`, `u[n + 2]`, and so forth, will contain $u_{n+1}^{m+1,k}$, $u_{n+2}^{m+1,k}$, etc. The algorithm is considered to have converged once the sum over n of the squares of the difference between $u_n^{m+1,k+1}$ and $u_n^{m+1,k}$ is less than the error tolerance `eps`. At this point the array `u[ ]` contains the SOR solution for u_n^{m+1}. Note that the routine solves the problem only for `Nminus` + 1 ≤ `n` ≤ `Nplus` − 1; the values at `Nplus` and `Nminus` are assumed to have been set by the calling routine. The routine returns the number of iterations executed in `loops`; this is so the calling routine may adjust `omega` to minimise the number of iterations.

Again, it can be shown that the Gauss–Seidel algorithm converges to the correct solution if α is positive, although we do not show it here.

The **SOR** algorithm is a refinement of the Gauss–Seidel method. We begin with the (seemingly trivial) observation that

$$u_n^{m+1,k+1} = u_n^{m+1,k} + (u_n^{m+1,k+1} - u_n^{m+1,k}).$$

As the sequence of iterates $u_n^{m+1,k}$ is intended to converge to u_n^{m+1} as $k \to \infty$, we can think of $(u_n^{m+1,k+1} - u_n^{m+1,k})$ as a correction term to be added to $u_n^{m+1,k}$ to bring it nearer to the exact value of u_n^{m+1}. The possibility then arises that we might be able to get the sequence to converge more rapidly if we over-correct; this is true if the sequence of iterates $u_n^{m+1,k} \to u_n^{m+1}$ monotonically as k increases, rather than oscillating; this is the case for both Gauss–Seidel and SOR. That is, we put

$$\begin{aligned} y_n^{m+1,k+1} &= \frac{1}{1+2\alpha}\left(b_n^m + \alpha(u_{n-1}^{m+1,k+1} + u_{n+1}^{m+1,k})\right) \\ u_n^{m+1,k+1} &= u_n^{m+1,k} + \omega(y_n^{m+1,k+1} - u_n^{m+1,k}), \end{aligned} \tag{8.27}$$

where $\omega > 1$ is the over-correction or **over-relaxation** parameter. (Note that the term $y_n^{m+1,k+1}$ is the value that the Gauss–Seidel method would give for $u_n^{m+1,k+1}$; in SOR we view $y_n^{m+1,k+1} - u_n^{m+1,k}$ as a correction to be made to $u_n^{m+1,k}$ in order to obtain $u_n^{m+1,k+1}$.) It can be shown that the SOR algorithm converges to the correct solution of (8.14) or (8.16) if $\alpha > 0$ and provided $0 < \omega < 2$. (When $0 < \omega < 1$ the algorithm is referred to as under-relaxation rather than over-relaxation, which is used for cases where $1 < \omega < 2$.) It can be shown that there is an optimal value of ω, in the interval $1 < \omega < 2$, which leads to much more rapid convergence than other values of ω. This optimal value of ω depends on the dimension of the matrix involved and, more generally, on the details of the matrix involved. There are means of calculating or estimating the optimal value of ω, but typically these involve so many calculations that it is quicker to change ω at each time-step until a value is found that minimises the number of iterations of the SOR loop. In Figure 8.9 we give the SOR algorithm for the fully implicit finite-difference equations for the diffusion equation.

8.6.4 The Implicit Finite-difference Algorithm

The implicit finite-difference solution scheme is to solve (8.16) (or (8.14) or (8.15)) for each time-step using either the LU solver routine described in Section (8.6.2) or the SOR algorithm described in the previous section. This allows us to time-step through and calculate the current value of the option. The algorithm using the LU method is illustrated in Figure 8.10, and the algorithm using the SOR method is illustrated in Figure 8.11.

In Figure 8.12 we compare implicit finite-difference solutions for a European put with a three month expiry time, exercise price $E = 10$, volatility $\sigma = 0.4$ and risk-free interest rate $r = 0.1$ with the exact Black–Scholes formula. As with the explicit method, the x-mesh spacing is first fixed and the time-step subsequently determined from α. We have chosen, as before, to regard α and $\delta\tau$ as variable to highlight an important point about stability. Whether $\alpha = 0.5$, $\alpha = 1.0$ or $\alpha = 5.0$,

```
implicit_fd1( values,dx,dt,M,Nminus,Nplus )
{
  a = dt/(dx*dx);

  for( n=Nminus; n<=Nplus; ++n )
    values[n] = pay_off(n*dx);

  lu_find_y( y,a,Nminus,Nplus );

  for( m=1; m<=M; ++m )
  {
    tau = m*dt;

    for( n=Nminus+1; n<Nplus; ++n )
      b[n] = values[n];

    values[Nminus] = u_m_inf( Nminus*dx, tau );
    values[ Nplus] = u_p_inf(  Nplus*dx, tau );
    b[Nminus+1] += a*values[Nminus];
    b[ Nplus-1] += a*values[ Nplus];

    lu_solver( values,b,y,a,Nminus,Nplus );
  }
}
```

Figure 8.10 Pseudo-code for an implicit solver using LU decomposition. `M` is the number of time-steps and `a` $= \alpha$. Note that we have to call **`lu_find_y`** once (and only once) before using the routine **`lu_solver`**. We then store the initial values in the array **`values`**[ ] and repeatedly call **`lu_solver`** to time-step through to expiry. Note the boundary value correction applied to the end values `b`[**`Nminus`** + **`1`**] and `b`[**`Nplus`** − **`1`**].

the computed solution agrees quite well with the exact solution and there is certainly no evidence of the sort of instability seen in the explicit finite-difference solution when $\alpha > \frac{1}{2}$.

This illustrates the fact that the implicit scheme is stable where the explicit scheme is unstable (that is, for $\alpha > \frac{1}{2}$). In fact, we can show that the implicit finite-difference method is stable for any $\alpha > 0$; see Exercise 11. The consequence is that we can solve the diffusion equation with larger time-steps using an implicit algorithm than we can using an explicit algorithm. This leads to more efficient numerical solutions; even though each time-step takes slightly longer in the implicit method the need for fewer time-steps more than compensates for this. The convergence of the implicit finite-difference approximation to the solution of

```
implicit_fd2( values,dx,dt,M,Nminus,Nplus )
{
  a = dt/(dx*dx);
  eps = 1.0e-8;
  omega = 1.0;
  domega = 0.05;
  oldloops = 10000;

  for( n=Nminus; n<=Nplus; ++n )
    values[n] = pay_off(n*dx);

  for( m=1; m<=M; ++m )
  {
    tau = m*dt;

    for( n=Nminus+1; n<Nplus; ++n )
      b[n] = values[n];

    values[Nminus] = u_m_inf( Nminus*dx, tau );
    values[ Nplus] = u_p_inf(  Nplus*dx, tau );

    SOR_solver( values,b,Nminus,Nplus,a,omega,eps,loops );
    if ( loops > oldloops ) domega *= -1.0;
    omega += domega;
    oldloops = loops;
  }
}
```

Figure 8.11 Pseudo-code for an implicit solver using the SOR method. `M` is the number of time-steps, `a` $= \alpha$, `eps` is the error tolerance and `omega` the SOR parameter. As in the previous pseudo-code, the routine first puts the initial values into the array `values[]` and then repeatedly calls `SOR_solver` to time-step through to expiry. There is no need to apply boundary value corrections to the end values `b[Nminus + 1]` and `b[Nplus − 1]`, as the SOR routine does this automatically. The routine increments `omega` by `domega` at each step. If this results in the number of iterations for the current step, `loops`, exceeding the iterations for the previous loop, `oldloops`, then the sign of `domega` is changed so that `omega` is moved back towards the value that minimises the number of iterations.

the partial differential equation can be proved. Again, like the explicit finite-difference scheme, it is convergent if and only if it is stable.

8.7 The Crank–Nicolson Method

The Crank–Nicolson finite-difference method is used to overcome the stability limitations imposed by the stability and convergence restrictions of the explicit finite-difference method, and to have $O((\delta\tau)^2)$ rates

S	$\alpha = 0.50$	$\alpha = 1.00$	$\alpha = 5.00$	exact
0.00	9.7531	9.7531	9.7531	9.7531
2.00	7.7531	7.7531	7.7531	7.7531
4.00	5.7531	5.7531	5.7530	5.7531
6.00	3.7569	3.7570	3.7573	3.7569
8.00	1.9025	1.9025	1.9030	1.9024
10.00	0.6690	0.6689	0.6675	0.6694
12.00	0.1674	0.1674	0.1670	0.1675
14.00	0.0327	0.0328	0.0332	0.0326
16.00	0.0054	0.0055	0.0058	0.0054

Figure 8.12 Comparison of exact Black–Scholes and fully implicit finite-difference solutions for a European put with $E = 10$, $r = 0.1$, $\sigma = 0.4$ and three months to expiry. Even with $\alpha = 5.0$ the results are accurate to 2 decimal places.

of convergence to the solution of the partial differential equation. (The rate of convergence of the implicit and explicit methods is $O(\delta\tau)$.)

The Crank–Nicolson implicit finite-difference scheme is essentially an average of the implicit and explicit methods. Specifically, if we use a forward-difference approximation to the time partial derivative we obtain the explicit scheme

$$\frac{u_n^{m+1} - u_n^m}{\delta\tau} + O(\delta\tau) = \frac{u_{n+1}^m - 2u_n^m + u_{n-1}^m}{(\delta x)^2} + O\left((\delta x)^2\right),$$

and if we take a backward difference we obtain the implicit scheme

$$\frac{u_n^{m+1} - u_n^m}{\delta\tau} + O(\delta\tau) = \frac{u_{n+1}^{m+1} - 2u_n^{m+1} + u_{n-1}^{m+1}}{(\delta x)^2} + O\left((\delta x)^2\right).$$

The average of these two equations is

$$\frac{u_n^{m+1} - u_n^m}{\delta\tau} + O(\delta\tau) =$$
$$\frac{1}{2}\left(\frac{u_{n+1}^m - 2u_n^m + u_{n-1}^m}{(\delta x)^2} + \frac{u_{n+1}^{m+1} - 2u_n^{m+1} + u_{n-1}^{m+1}}{(\delta x)^2}\right) + O\left((\delta x)^2\right). \tag{8.28}$$

In fact, it can be shown that the terms in (8.28) are accurate to $O((\delta\tau)^2)$, rather than $O(\delta\tau)$; see Exercise 16. Ignoring the error terms leads to the Crank–Nicolson scheme

$$\begin{aligned} u_n^{m+1} - \tfrac{1}{2}\alpha(u_{n-1}^{m+1} - 2u_n^{m+1} + u_{n+1}^{m+1}) \\ = u_n^m + \tfrac{1}{2}\alpha(u_{n-1}^m - 2u_n^m + u_{n+1}^m) \end{aligned} \tag{8.29}$$

where, as before, $\alpha = \delta\tau/(\delta x)^2$. Note that u_n^{m+1}, u_{n-1}^{m+1} and u_{n+1}^{m+1} are now determined implicitly in terms of all of u_n^m, u_{n+1}^m and u_{n-1}^m.

Solving this system of equations is, in principle, no different from solving the equations (8.14) for the implicit scheme. This is because everything on the right-hand side of (8.29) can be evaluated explicitly if the u_n^m are known. The problem thus reduces to first calculating

$$Z_n^m = (1-\alpha)u_n^m + \tfrac{1}{2}\alpha(u_{n-1}^m + u_{n+1}^m), \tag{8.30}$$

which is an explicit formula for Z_n^m, and then solving

$$(1+\alpha)u_n^{m+1} - \tfrac{1}{2}\alpha(u_{n-1}^{m+1} + u_{n+1}^{m+1}) = Z_n^m. \tag{8.31}$$

This second problem is essentially the same as solving (8.14).

Again we assume that we can truncate the infinite mesh at $x = N^-\delta x$ and $x = N^+\delta x$, and take N^- and N^+ sufficiently large so that no significant errors are introduced. As before we can calculate u_n^0 using (8.13) and $u_{N^-}^{m+1}$ and $u_{N^+}^{m+1}$ from the boundary conditions (8.12).

There remains the problem of finding the u_n^{m+1} for $m \geq 0$ and $N^- < n < N^+$ from (8.31). We can write the problem as a linear system

$$\mathbf{C}\boldsymbol{u}^{m+1} = \boldsymbol{b}^m \tag{8.32}$$

where the matrix $\mathbf{C}$ is given by

$$\mathbf{C} = \begin{pmatrix} 1+\alpha & -\frac{1}{2}\alpha & 0 & \cdots & 0 \\ -\frac{1}{2}\alpha & 1+\alpha & -\frac{1}{2}\alpha & & \vdots \\ 0 & -\frac{1}{2}\alpha & \ddots & \ddots & 0 \\ \vdots & & \ddots & \ddots & -\frac{1}{2}\alpha \\ 0 & 0 & & -\frac{1}{2}\alpha & 1+\alpha \end{pmatrix}, \tag{8.33}$$

and the vectors $\boldsymbol{u}^{m+1}$ and $\boldsymbol{b}^m$ are given by

$$\boldsymbol{u}^{m+1} = \begin{pmatrix} u_{N^-+1}^{m+1} \\ \vdots \\ u_0^{m+1} \\ \vdots \\ u_{N^+-1}^{m+1} \end{pmatrix}, \quad \boldsymbol{b}^m = \begin{pmatrix} Z_{N^-+1}^m \\ \vdots \\ Z_0^m \\ \vdots \\ Z_{N^+-1}^m \end{pmatrix} + \tfrac{1}{2}\alpha \begin{pmatrix} u_{N^-}^{m+1} \\ 0 \\ \vdots \\ 0 \\ u_{N^+}^{m+1} \end{pmatrix}. \tag{8.34}$$

The vector on the extreme right-hand side of equation (8.34), in $\boldsymbol{b}^m$, arises from the boundary conditions applied at the ends, as in the fully implicit finite-difference method.

```
crank_fd1( val,dx,dt,M,Nminus,Nplus )
{
  a = dt/(dx*dx);
  a2 = a/2.0;

  for( n=Nminus; n<=Nplus; ++n )
    val[n] = pay_off( n*dx );

  lu_find_y( y,a2,Nminus,Nplus );

  for( m=1; m<=M; ++m )
  {
    tau = m*dt;
    for( n=Nminus+1; n<Nplus; ++n )
      b[n] = (1-a)*val[n]+a2*(val[n+1]+val[n-1]);

    val[Nminus] = u_m_inf( Nminus*dx,tau );
    val[ Nplus] = u_p_inf(  Nplus*dx,tau );
    b[Nminus+1] += a2*val[Nminus];
    b[ Nplus-1] += a2*val[ Nplus];

    lu_solver( val,b,y,a2,Nminus,Nplus );
  }
}
```

Figure 8.13 Pseudo-code for a Crank–Nicolson solver using the LU method. The solution after `M` time-steps is returned in the array `val[]`. Here $\mathtt{a} = \alpha$ and $\mathtt{a2} = \alpha/2$. Note the boundary corrections to `b[Nminus + 1]` and `b[Nplus − 1]`. Note also that the code is almost identical to the code in Figure 8.10; the main differences are in the use of $\alpha/2$ instead of α and in the calculation of `b[n]`.

To implement the Crank–Nicolson scheme, we first form the vector $\boldsymbol{b}^m$ using known quantities. Then we use an LU tridiagonal solver or an SOR solver to solve the system (8.32). This allows us to time-step through the solution. The only difference between the LU or SOR solvers for the Crank–Nicolson and fully implicit methods is that whenever α appears in the algorithm for the implicit scheme, we replace it by $\frac{1}{2}\alpha$ for the Crank–Nicolson scheme. In Figures 8.13 and 8.14 we give pseudo-codes for the Crank–Nicolson method using LU and SOR solution methods respectively.

In Figure 8.15 we compare Crank–Nicolson finite-difference solutions for a European put with four months to expiry, exercise price 10, volatility $\sigma = 0.45$ and risk-free interest rate $r = 0.1$ with the exact Black–Scholes formula. Notice that with $\alpha = \frac{1}{2}$, $\alpha = 1.0$ and even $\alpha = 10.0$,

```
crank_fd2( val,dx,dt,M, Nminus,Nplus )
{
  a = dt/(dx*dx);
  a2 = a/2.0;
  eps = 1.0e-8;
  omega = 1.0;
  domega = 0.05;
  oldloops = 10000;

  for( n=Nminus; n<=Nplus; ++n )
    val[n] = pay_off( n*dx );

  for( m=1; m<=M; ++m )
  {
    tau = m*dt;

    for( n=Nminus+1; n<Nplus; ++n )
      b[n] = (1-a)*val[n]+a2*(val[n+1]+val[n-1]);

    val[Nminus] = u_m_inf(Nminus*dx,tau);
    val[ Nplus] = u_p_inf( Nplus*dx,tau);

    SOR_solver( val,b,Nminus,Nplus,a2,omega,eps,loops );
    if ( loops > oldloops ) domega *= -1.0;
    omega += domega;
    oldloops = loops;
  }
}
```

Figure 8.14 Pseudo-code for a Crank–Nicolson solver using the SOR method. The solution after `M` time-steps is returned in the array `val[]`. Here `a` $= \alpha$ and `a2` $= \alpha/2$. Note the boundary corrections to `b[Nminus + 1]` and `b[Nplus − 1]`. Note also that the code is almost identical to the code in Figure 8.11; the main differences are in the use of $\alpha/2$ instead of α and in the calculation of `b[n]`.

the computed solution agrees very well with the exact solution. This demonstrates the fact that the Crank–Nicolson scheme is stable where the explicit scheme is unstable (that is, for $\alpha > \frac{1}{2}$). Moreover, its accuracy is greater than that of the fully implicit scheme.

We can show that the Crank–Nicolson scheme is both stable and convergent for all values of $\alpha > 0$. For the proof of stability see Exercise 17.

S	$\alpha = 0.50$	$\alpha = 1.00$	$\alpha = 10.00$	exact
0.00	9.6722	9.6722	9.6722	9.6722
2.00	7.6721	7.6721	7.6721	7.6722
4.00	5.6722	5.6722	5.6723	5.6723
6.00	3.6976	3.6976	3.6975	3.6977
8.00	1.9804	1.9804	1.9804	1.9806
10.00	0.8605	0.8605	0.8566	0.8610
12.00	0.3174	0.3174	0.3174	0.3174
14.00	0.1047	0.1047	0.1046	0.1046
16.00	0.0322	0.0322	0.0321	0.0322

Figure 8.15 Comparison of exact Black–Scholes and Crank–Nicolson finite-difference solutions for a European put with $E = 10$, $r = 0.1$, $\sigma = 0.45$ and four months to expiry. Even with $\alpha = 10$, the numerical and exact results differ only marginally.

Further Reading

- The books by Johnson & Riess (1982), Strang (1986) and Stoer & Bulirsch (1993) give excellent background material on some of the basic considerations of numerical analysis: accuracy of computer arithmetic, convergence and efficiency of algorithms and stability of numerical methods.
- Richtmyer & Morton (1967) and Smith (1985) are both very readable books on the subject of finite-difference methods.
- Brennan & Schwartz (1978) were the first to describe the application of finite-difference methods to option pricing.
- Geske & Shastri (1985) give a comparison of the efficiency of various finite-difference and other numerical methods for option pricing.

Exercises

1. By considering the Taylor series expansion of $u(x, \tau + \delta\tau)$ about (x, τ), show that the forward difference (8.1) satisfies

$$\frac{u(x, \tau + \delta\tau) - u(x, \tau)}{\delta\tau} = \frac{\partial u}{\partial \tau}(x, \tau) + \frac{\partial^2 u}{\partial \tau^2}(x, \tau + \lambda\,\delta\tau)\,\delta\tau$$

for some $0 \le \lambda \le 1$. Obtain a similar result for the backward-difference approximation (8.2)

2. Expand $u(x, \tau + \delta\tau)$ and $u(x, \tau - \delta\tau)$ as Taylor series about (x, τ). Deduce that the central-difference approximations (8.3) and (8.4) are indeed accurate to $O\Big((\delta\tau)^2\Big)$.

3. Show that the central-symmetric-difference approximation (8.6) is accurate to $O\Big((\delta x)^2\Big)$.

4. Find the minimum number of arithmetical operations (divisions and multiplications) required per time-step of the explicit algorithm (8.10).

5. **Stability I.** Suppose that e_n^m are the finite-precision errors introduced into the solution of (8.10) because of initial rounding errors, e_n^0, in the exact initial values when represented on a computer. Why do the e_n^m also satisfy (8.10)? As a consequence of Fourier analysis, we may assume (without any loss of generality) that the errors take the form $e_n^m = \lambda^m \sin(n\omega)$ for some given frequency ω. Deduce from (8.10) that

$$\lambda = 1 - 4\alpha \sin^2(\tfrac{1}{2}\omega)$$

and infer that unless $0 \leq \alpha \leq \frac{1}{2}$ there are always frequencies ω for which the error grows without bound.

6. For the explicit finite-difference scheme, show that if we increase the number of x-points by a factor of K, then the number of calculations performed increases by a factor of K^3 (assuming α remains constant).

7. What changes would be necessary to the boundary and initial conditions for the explicit finite-difference method to value a bullish vertical spread? What modifications would be needed to the pseudo-code in Figure 8.5 ?

8. Write a computer program that implements the explicit finite-difference method for calls, puts, cash-or-nothing calls and cash-or-nothing puts.

9. Consider the untransformed Black–Scholes equation (6.3). Use a mesh of equal S-steps of size δS and equal time-steps of size δt, central differences for S derivatives and backward differences for time derivatives to obtain the explicit finite-difference equations

$$V_n^m = a_n V_{n-1}^{m+1} + b_n V_n^{m+1} + c_n V_{n+1}^{m+1}, \quad n = 0, 1, 2, 3, \ldots$$

where V_n^m is the finite-difference approximation to $V(n\,\delta S, m\,\delta t)$ and

$$\begin{aligned} a_n &= \tfrac{1}{2}\left(\sigma^2 n^2 - (r - D_0)n\right)\delta t \\ b_n &= 1 - \left(\sigma^2 n^2 + r\right)\delta t \\ c_n &= \tfrac{1}{2}\left(\sigma^2 n^2 + (r - D_0)n\right)\delta t. \end{aligned}$$

Why is this an *explicit* method? What boundary and initial or final conditions are appropriate? What stability problems can you see arising?

10. Find the minimum number of arithmetical operations (divisions and multiplications) required per time-step of the implicit algorithm (8.14) assuming that an LU solver is used.

11. **Stability II.** Suppose that e_n^m are the finite-precision arithmetical errors introduced into the solution of (8.14) because of initial rounding errors, e_n^0, in the exact initial values when represented on a computer. Following the analysis of Exercise 5, show that if the errors take the form $e_n^m = \lambda^m \sin(n\omega)$ for some given frequency ω then, from (8.14),

$$\lambda = \frac{1}{1 + 4\alpha \sin^2(\frac{1}{2}\omega)}.$$

Deduce that if $\alpha > 0$ errors of any frequency ω do not grow in time.

12. Write a computer program that implements the fully implicit finite-difference method for calls, puts, cash-or-nothing calls and puts.

13. With the same notation as in Exercise 9 show that the implicit discretisation of the Black–Scholes equation may be written as

$$B_0 V_0^m + C_0 V_1^m = V_0^{m+1},$$

$$A_n V_{n-1}^m + B_n V_n^m + C_n V_{n+1}^m = V_n^{m+1}, \quad n = 1, 2, 3, \ldots, N,$$

where

$$\begin{aligned} A_n &= -\tfrac{1}{2}\left(\sigma^2 n^2 - (r - D_0)n\right)\delta t \\ B_n &= 1 + \left(\sigma^2 n^2 + r\right)\delta t \\ C_n &= -\tfrac{1}{2}\left(\sigma^2 n^2 + (r - D_0)n\right)\delta t. \end{aligned}$$

14. Show that LU decomposition for the system

$$\begin{pmatrix} B_0 & C_0 & 0 & \cdots & 0 \\ A_1 & B_1 & C_1 & & \vdots \\ 0 & A_2 & \ddots & \ddots & 0 \\ \vdots & & \ddots & \ddots & C_{N-1} \\ 0 & \cdots & 0 & A_N & B_N \end{pmatrix} \begin{pmatrix} V_0^m \\ V_1^m \\ \vdots \\ V_{N-1}^m \\ V_N^m \end{pmatrix} = \begin{pmatrix} V_0^{m+1} \\ V_1^{m+1} \\ \vdots \\ V_{N-1}^{m+1} \\ V_N^{m+1} \end{pmatrix},$$

of Exercise 13, leads to the algorithm

$$\begin{aligned} q_0^m &= V_0^{m+1}, & q_n^m &= V_n^{m+1} - A_n q_{n-1}^m / F_{n-1}, \quad n = 1, 2, \ldots, N \\ V_N^m &= q_N^m / F_N, & V_n^m &= (q_n^m - C_n V_{n+1}^m)/F_n, \quad n = N-1, \ldots, 2, 1, \end{aligned}$$

where the q_n are intermediate quantities and the F_n are calculated from the matrix above using

$$F_0 = B_0, \qquad F_n = B_n - C_{n-1}A_n/F_{n-1}, \quad n = 1, 2, \ldots, N.$$

15. Describe the Jacobi, Gauss-Seidel and SOR algorithms for the system in Exercises 13 and 14.

16. Show that the right-hand side of (8.28) is a finite-difference approximation to $\frac{1}{2}\left(\partial^2 u/\partial x^2(x, \tau + \delta\tau) + \partial^2 u/\partial x^2(x, \tau)\right)$. By using Taylor's theorem deduce that

$$\frac{\partial^2 u}{\partial x^2}(x, \tau + \delta\tau/2) = \frac{1}{2}\left(\frac{\partial^2 u}{\partial x^2}(x, \tau) + \frac{\partial^2 u}{\partial x^2}(x, \tau + \delta\tau)\right) + O\left((\delta\tau)^2\right).$$

Use Exercise 2 to show that

$$\frac{\partial u}{\partial \tau}(x, \tau + \delta\tau/2) = \frac{u(x, \tau + \delta\tau) - u(x, \tau)}{\delta\tau} + O\left((\delta\tau)^2\right).$$

Deduce that (8.28), viewed as a finite-difference approximation to

$$\frac{\partial u}{\partial \tau}(x, \tau + \delta\tau/2) = \frac{\partial^2 u}{\partial x^2}(x, \tau + \delta\tau/2),$$

is accurate to $O\Big((\delta\tau)^2 + (\delta x)^2\Big)$.

17. **Stability III.** Repeat the stability calculations in Exercises 5 and 11, but in this case for the Crank–Nicolson equations (8.31). Show that an error term $e_n^m = \lambda^m \sin(n\omega)$ implies

$$\lambda = \frac{1 - 2\alpha \sin^2 \frac{1}{2}\omega}{1 + 2\alpha \sin^2 \frac{1}{2}\omega}.$$

Deduce that the Crank–Nicolson scheme is stable for all $\alpha > 0$.

18. Write down the system of equations resulting from the direct application of the Crank–Nicolson method to the Black–Scholes equation. Describe the LU decomposition and SOR algorithms for its solution.

19. Consider the diffusion equation problem (this problem arises, for example, from an option whose payoff depends on the difference in prices of

two uncorrelated assets)

$$
\begin{aligned}
\frac{\partial u}{\partial \tau} &= \frac{\partial^2 u}{\partial x^2} + \frac{\partial^2 u}{\partial y^2}, \\
u(x,y,0) &= u_0(x,y), \\
\text{as } x \to -\infty, \quad u(x,y,\tau) &\sim u^1_{-\infty}(y,\tau), \\
\text{as } x \to \infty, \quad u(x,y,\tau) &\sim u^1_{\infty}(y,\tau), \\
\text{as } y \to -\infty, \quad u(x,y,\tau) &\sim u^2_{-\infty}(x,\tau), \\
\text{as } y \to \infty, \quad u(x,y,\tau) &\sim u^2_{\infty}(x,\tau).
\end{aligned}
$$

Assume equal x- and y-step sizes, $\delta x = \delta y$, a square grid $N^-\,\delta x \le i\,\delta x \le N^+\,\delta x$ and $N^-\,\delta y \le j\,\delta y \le N^+\,\delta y$, and let u^m_{ij} denote the finite-difference approximation to $u(i\,\delta x, j\,\delta y, m\,\delta\tau)$. Write down the explicit, fully implicit and Crank–Nicolson finite-difference schemes for this problem. What stability restrictions apply to the explicit method?

9 Methods for American Options

9.1 Introduction

Using finite-difference methods for European options is relatively straightforward, as there is no possibility of early exercise. As we have seen, the possibility of early exercise may lead to free boundaries. The chief problem with free boundaries, from the point of view of numerical analysis, is that we do not know where they are. This makes it difficult to impose the free boundary conditions, since we have to determine where to impose them as part of the solution procedure. (Recall that in Chapter 8 we simply imposed the boundary conditions at fixed grid points.)

There are two distinct strategies for the numerical solution of free boundary problems. One is to attempt to track the free boundary as part of the time-stepping process. In the context of valuation of American options this is not a particularly attractive method, as the free boundary conditions are both implicit – that is, they do not give a direct expression for the free boundary or its time derivatives. We simply note the existence of such methods here, and refer the reader to the literature for a discussion of various boundary tracking strategies for implicit free boundary problems.

The other strategy is to attempt to find a transformation that reduces the problem to a fixed boundary problem from which the free boundary can be inferred *afterwards*. There are many transformations that do this, but we consider only the particularly elegant method involving the use of the linear complementarity formulation.

Recall from Chapter 7 that we can write the American option valuation problem in the compact linear complementarity form

$$\begin{gathered}\left(\frac{\partial u}{\partial \tau}-\frac{\partial^2 u}{\partial x^2}\right)\geq 0, \quad \Big(u(x,\tau)-g(x,\tau)\Big)\geq 0,\\ \left(\frac{\partial u}{\partial \tau}-\frac{\partial^2 u}{\partial x^2}\right)\cdot\Big(u(x,\tau)-g(x,\tau)\Big)=0,\end{gathered} \tag{9.1}$$

where, as before, we are using the change of variables (5.9). The transformed payoff constraint function, $g(x,\tau)$, is given by

$$g(x,\tau)=e^{\frac{1}{4}(k+1)^2\tau}\ \max\left(e^{\frac{1}{2}(k-1)x}-e^{\frac{1}{2}(k+1)x},0\right)$$

for the put,

$$g(x,\tau)=e^{\frac{1}{4}(k+1)^2\tau}\ \max\left(e^{\frac{1}{2}(k+1)x}-e^{\frac{1}{2}(k-1)x},0\right)$$

for the call and

$$g(x,\tau)=e^{\frac{1}{4}(k+1)^2\tau}\times\begin{cases}0 & x<0\\ be^{\frac{1}{2}(k-1)x} & x\geq 0\end{cases}$$

for the cash-or-nothing call and where $k=r/\frac{1}{2}\sigma^2$. The initial and fixed boundary conditions become

$$\begin{gathered}u(x,0)=g(x,0),\\ u(x,\tau)\ \text{is continuous},\\ \frac{\partial u}{\partial x}(x,\tau)\ \text{is as continuous as}\ g(x,\tau),\\ \lim_{x\to\pm\infty}u(x,\tau)=\lim_{x\to\pm\infty}g(x,\tau).\end{gathered} \tag{9.2}$$

This framework extends in the obvious way to more general payoff functions than the three forms for $g(x,\tau)$ given above.

As noted already, one of the main advantages of the linear complementarity formulation (9.1) is that there is no explicit mention of the free boundary. If we can solve the linear complementarity problem then we find the free boundary $x=x_f(\tau)$ *a posteriori* by the condition that defines it, namely that

$$u\big(x_f(\tau),\tau\big)=g\big(x_f(\tau),\tau\big),\quad \text{but}\quad u(x,\tau)>g(x,\tau)\ \text{for}\ x>x_f(\tau)$$

for the put, and

$$u\big(x_f(\tau),\tau\big)=g\big(x_f(\tau),\tau\big),\quad \text{but}\quad u(x,\tau)>g(x,\tau)\ \text{for}\ x<x_f(\tau)$$

for the calls. The formulation remains valid if there are several free boundaries, or indeed if there are none at all; the free boundaries are defined as the points where $u(x,\tau)$ first meets $g(x,\tau)$.

9.2 Finite-difference Formulation

We divide the (x, τ)-plane into a regular finite mesh as usual, and take a finite-difference approximation of the linear complementarity equations (9.1). As most of this discretisation is a simple extension of the finite-difference formulations given in the previous chapter, we only give a short account of it here.

We approximate terms of the form $\partial u/\partial \tau - \partial^2 u/\partial x^2$ by finite differences on a regular mesh with step sizes $\delta\tau$ and δx, and truncating so that x lies between $N^- \,\delta x$ and $N^+ \,\delta x$,

$$N^- \,\delta x \le x = n\,\delta x \le N^+ \,\delta x,$$

where $-N^-$ and N^+ are suitably large numbers.

Rather than going through the cases of explicit, implicit and Crank–Nicolson methods separately, we only consider the Crank–Nicolson method in any detail, and leave the other formulations as exercises. Thus we take (see Exercise 16 of the previous chapter)

$$\frac{\partial u}{\partial \tau}(x, \tau + \delta\tau/2) = \frac{u_n^{m+1} - u_n^m}{\delta\tau} + O\Big((\delta\tau)^2\Big),$$

and

$$\begin{aligned}\frac{\partial^2 u}{\partial x^2}(x, \tau + \delta\tau/2) = {} & \frac{1}{2}\left(\frac{u_{n+1}^{m+1} - 2u_n^{m+1} + u_{n-1}^{m+1}}{(\delta x)^2}\right) \\ & + \frac{1}{2}\left(\frac{u_{n+1}^{m} - 2u_n^{m} + u_{n-1}^{m}}{(\delta x)^2}\right) + O\Big((\delta x)^2\Big),\end{aligned}$$

where, as usual, $u_n^m = u(n\,\delta x, m\,\delta\tau)$. Dropping terms of $O\Big((\delta\tau)^2\Big)$ and $O\Big((\delta x)^2\Big)$, the inequality $\partial u/\partial\tau - \partial^2 u/\partial x^2 \ge 0$ is approximated by

$$\begin{aligned}u_n^{m+1} - \tfrac{1}{2}\alpha(u_{n+1}^{m+1} - 2u_n^{m+1} + u_{n-1}^{m+1}) & \\ \ge u_n^m + \tfrac{1}{2}\alpha(u_{n+1}^{m} - 2u_n^{m} + u_{n-1}^{m}), &\end{aligned} \tag{9.3}$$

where, as usual, α is given by (8.11).

We write

$$g_n^m = g(n\,\delta x, m\,\delta\tau) \tag{9.4}$$

for the discretised payoff function. The condition $u(x, \tau) \ge g(x, \tau)$ is approximated by

$$u_n^m \ge g_n^m \quad \text{for } m \ge 1, \tag{9.5}$$

and the boundary and initial conditions (9.2) imply that

$$u_{N^-}^m = g_{N^-}^m, \quad u_{N^+}^m = g_{N^+}^m, \quad u_n^0 = g_n^0. \tag{9.6}$$

If we define Z_n^m by

$$Z_n^m = (1-\alpha)u_n^m + \tfrac{1}{2}\alpha(u_{n+1}^m + u_{n-1}^m), \tag{9.7}$$

then (9.3) becomes

$$(1+\alpha)u_n^{m+1} - \tfrac{1}{2}\alpha(u_{n+1}^{m+1} + u_{n-1}^{m+1}) \geq Z_n^m. \tag{9.8}$$

Note that at time-step $(m+1)\delta\tau$ we can find Z_n^m explicitly, since we know the values of u_n^m. The linear complementarity condition that

$$\left(\frac{\partial u}{\partial \tau} - \frac{\partial^2 u}{\partial x^2}\right) \cdot \Big(u(x,\tau) - g(x,\tau)\Big) = 0$$

is approximated by

$$\left((1+\alpha)u_n^{m+1} - \tfrac{1}{2}\alpha(u_{n+1}^{m+1} + u_{n-1}^{m+1}) - Z_n^m\right)\left(u_n^{m+1} - g_n^{m+1}\right) = 0. \tag{9.9}$$

9.3 The Constrained Matrix Problem

We now formulate the finite-difference approximation (9.5)–(9.9) as a constrained matrix problem and, in the following section describe how to solve this constrained problem using the **projected SOR** method.

Let $\boldsymbol{u}^m$ denote the vector of approximate values at time-step $m\,\delta\tau$ and $\boldsymbol{g}^m$ the vector representing the constraint at this same time:

$$\boldsymbol{u}^m = \begin{pmatrix} u_{N^-+1}^m \\ \vdots \\ u_{N^+-1}^m \end{pmatrix}, \quad \boldsymbol{g}^m = \begin{pmatrix} g_{N^-+1}^m \\ \vdots \\ g_{N^+-1}^m \end{pmatrix}. \tag{9.10}$$

We do not include the terms $u_{N^+}^m$ and $u_{N^+}^m$, as they are explicitly determined by the boundary conditions (9.6). Let the vector $\boldsymbol{b}^m$ be given by

$$\boldsymbol{b}^m = \begin{pmatrix} b_{N^-+1}^m \\ \vdots \\ b_0^m \\ \vdots \\ b_{N^+-1}^m \end{pmatrix} = \begin{pmatrix} Z_{N^-+1}^m \\ \vdots \\ Z_0^m \\ \vdots \\ Z_{N^+-1}^m \end{pmatrix} + \tfrac{1}{2}\alpha \begin{pmatrix} g_{N^-}^{m+1} \\ 0 \\ \vdots \\ 0 \\ g_{N^+}^{m+1} \end{pmatrix}, \tag{9.11}$$

where the quantities Z_n^m are determined from (9.7) and the vector on the extreme right-hand side of (9.11) includes the effects of the boundary conditions at $n = N^-$ and $n = N^+$.

If we reintroduce the $(N^+ - N^- - 1)$-square, tridiagonal, symmetric matrix (which we used in the Crank–Nicolson scheme in Chapter 8)

$$\mathbf{C} = \begin{pmatrix} 1+\alpha & -\frac{1}{2}\alpha & 0 & \cdots & 0 \\ -\frac{1}{2}\alpha & 1+\alpha & -\frac{1}{2}\alpha & & \vdots \\ 0 & -\frac{1}{2}\alpha & \ddots & \ddots & 0 \\ \vdots & & \ddots & 1+\alpha & -\frac{1}{2}\alpha \\ 0 & \cdots & 0 & -\frac{1}{2}\alpha & 1+\alpha \end{pmatrix}, \tag{9.12}$$

we can rewrite our discrete approximation (9.5)–(9.9) to the linear complementarity problem (9.1)–(9.2) in matrix form, as

$$\begin{gathered} \mathbf{C}\boldsymbol{u}^{m+1} \geq \boldsymbol{b}^m, \quad \boldsymbol{u}^{m+1} \geq \boldsymbol{g}^{m+1}, \\ \left(\boldsymbol{u}^{m+1} - \boldsymbol{g}^{m+1}\right) \cdot \left(\mathbf{C}\boldsymbol{u}^{m+1} - \boldsymbol{b}^m\right) = 0. \end{gathered} \tag{9.13}$$

We take the expression $\boldsymbol{a} \geq \boldsymbol{b}$, where $\boldsymbol{a}$ and $\boldsymbol{b}$ are vectors, to mean that each component of $\boldsymbol{a}$ is greater than or equal to the corresponding component of $\boldsymbol{b}$, that is, $a_n \geq b_n$ for all n.

The time-stepping is immanent in the scheme: the vector $\boldsymbol{b}^m$ contains the information from the time-step $m\,\delta\tau$ that determines the value of $\boldsymbol{v}^{m+1}$ at time-step $(m+1)\delta\tau$. At each time-step we can calculate $\boldsymbol{b}^m$ from already known values of $\boldsymbol{v}^m$. We can calculate $\boldsymbol{g}^m$ for any $m\,\delta\tau$, and thus to time-step the system we need only solve the problem (9.13). This can be done using a modified form of SOR, known as **projected SOR**.

9.4 Projected SOR

Projected SOR is a minor modification of the SOR method described in Section 8.6.3. Consider the SOR algorithm (8.27). If we adapt this for a Crank–Nicolson finite-difference formulation of the problem we obtain the equations

$$\begin{aligned} y_n^{m+1,k+1} &= \frac{1}{1+\alpha}\left(b_n^m + \tfrac{1}{2}\alpha(u_{n-1}^{m+1,k+1} + u_{n+1}^{m+1,k})\right) \\ u_n^{m+1,k+1} &= u_n^{m+1,k} + \omega(y_n^{m+1,k+1} - u_n^{m+1,k}). \end{aligned}$$

If these equations are iterated until the $u_n^{m+1,k}$ converge to u_n^{m+1}, they give the solution of $\mathbf{C}\boldsymbol{u}^{m+1} = \boldsymbol{b}^m$. To satisfy the constraint that

```
PSOR_solver( u,b,g,Nminus,Nplus,a,omega,eps,loops )
{
  loops = 0;
  a2 = a/2.0;
  u[Nminus] = g[Nminus];
  u[ Nplus] = g[ Nplus];

  do
  {
    error = 0.0;

    for( n=Nminus+1; n<Nplus; ++n )
    {
      y = (b[n]+a2*(u[n-1]+u[n+1])/(1+a);
      y = max( g[n], u[n]+omega*(y-u[n]) );
      error += (u[n]-y)*(u[n]-y);
      u[n] = y;
    }
    ++loops;
  } while ( error > eps );
  return(loops);
}
```

Figure 9.1 Pseudo-code for the projected SOR algorithm for an American option problem using Crank–Nicolson finite differences. Here a $= \alpha$, a2 $= \frac{1}{2}\alpha$, Nplus $= N^+$, Nminus $= N^-$, u[n] $= u_n^{m+1,k}$ or $u_n^{m+1,k+1}$, b[n] $= b_n^m$, g[n] $= g_n^{m+1}$, y $= y_n^{m+1,k+1}$, omega $= \omega$ and eps is the desired error tolerance. The routine counts the number of iterations required and returns it in loops so that the calling program can optimize the over-relaxation parameter omega. Compare with Figure 8.9.

$\boldsymbol{u}^{m+1} \geq \boldsymbol{g}^{m+1}$ we simply modify the second of these equations to enforce this result:

$$u_n^{m+1,k+1} = \max\left(u_n^{m+1,k} + \omega(y_n^{m+1,k+1} - u_n^{m+1,k}), g_n^{m+1}\right).$$

Notice that the constraint is enforced at the same time as the iterate $u_n^{m+1,k+1}$ is calculated; the effect of the constraint is immediately felt in the calculation of $u_{n+1}^{m+1,k+1}$, $u_{n+2}^{m+1,k+1}$, etc. Thus the projected SOR algorithm is to iterate (on k) the equations

$$\begin{aligned} y_n^{m+1,k+1} &= \frac{1}{1+\alpha}\left(b_n^m + \tfrac{1}{2}\alpha(u_{n-1}^{m+1,k+1} + u_{n+1}^{m+1,k})\right) \\ u_n^{m+1,k+1} &= \max\left(u_n^{m+1,k} + \omega(y_n^{m+1,k+1} - u_n^{m+1,k}), g_n^{m+1}\right) \end{aligned} \tag{9.14}$$

until the difference $\|\boldsymbol{u}^{m+1,k+1} - \boldsymbol{u}^{m+1,k}\|$ is small enough to be regarded as negligible. One then puts $\boldsymbol{u}^{m+1} = \boldsymbol{u}^{m+1,k+1}$.

As the algorithm is iterative, any solution generated is self-consistent in that it does not violate any of the constraints (see the Technical Point below). Moreover, such a solution has the property that either $u_n^{m+1} = g_n^{m+1}$ or the n-th component of $\mathbf{C}\boldsymbol{u}^{m+1} - \boldsymbol{b}^m$ vanishes. Thus the algorithm guarantees that both $\boldsymbol{u}^{m+1} \geq \boldsymbol{g}^{m+1}$ and $(\mathbf{C}\boldsymbol{u}^{m+1} - \boldsymbol{b}^m) \cdot (\boldsymbol{u}^{m+1} - \boldsymbol{b}^m) = 0$. The condition that $\mathbf{C}\boldsymbol{u}^{m+1} \geq \boldsymbol{b}^m$ follows as a consequence of the structure of $\mathbf{C}$ (specifically, from the fact that it is positive definite; we refer the reader to the literature for details). The algorithm is illustrated in Figure 9.1 and discussed further in the following section.

Technical Point: The Need for Projected SOR.
Inspection of (9.13) shows that for each component u_n^{m+1} of the vector $\boldsymbol{u}^{m+1}$ there are two, and only two, possibilities:

(1) $u_n^{m+1} > g_n^{m+1}$;
(2) $u_n^{m+1} = g_n^{m+1}$.

Case 1 corresponds to the optimality of holding the option, and case 2 to the optimality of exercising the option. Further, from the linear complementarity condition in (9.13) we see that, for cases 1 and 2 respectively, we must have:

(1) $(\mathbf{C}\boldsymbol{u}^{m+1})_n = b_n^m$;
(2) $(\mathbf{C}\boldsymbol{u}^{m+1})_n > b_n^m$.

Notice that there is an internal consistency requirement here. It is not enough simply to solve $\mathbf{C}\boldsymbol{u}^{m+1} = \boldsymbol{b}^m$ and then apply the cut-off condition that any component u_n^{m+1} of $\boldsymbol{u}^{m+1}$ that is less than the corresponding component g_n^{m+1} of $\boldsymbol{g}^{m+1}$ is replaced by that latter component. This strategy is frequently used (and is, in fact, valid for the explicit finite-difference discretisation of the linear complementarity formulation of the problem – the projected SOR algorithm reduces to this strategy in the case of explicit differences). The reason that it is invalid for implicit finite-difference formulations of the problem is simply that the components u_n^{m+1} of $\boldsymbol{u}^{m+1}$ are all implicitly related to each other and we cannot modify one in isolation without affecting the others. If we modify $\boldsymbol{u}^{m+1}$ in the way suggested above, there are no guarantees that $\mathbf{C}\boldsymbol{u}^{m+1} \geq \boldsymbol{b}^m$ or that $(\mathbf{C}\boldsymbol{u}^{m+1} - \boldsymbol{b}^m) \cdot (\boldsymbol{u}^{m+1} - \boldsymbol{g}^{m+1}) = 0$. The effect is to produce spurious 'solutions' that either fail to meet the free boundary conditions (i.e. produce sub-optimal values for the option) or fail to satisfy the Black–Scholes inequality (i.e. produce values for which arbitrage opportunities are present). The projected-SOR algorithm is constructed in such a way as to guarantee an internally consistent solution of (9.13) (that is, one that satisfies all the constraints simultaneously). This internally consistent solution is, it turns out, also the unique solution of (9.13), although we do not give a proof of this in this book.

9.5 The Time-stepping Algorithm

The algorithm for time-stepping from $\boldsymbol{u}^m$ to $\boldsymbol{u}^{m+1}$ is thus a simple modification to the SOR solution for European options. Specifically, as above, let the vector

$$\boldsymbol{u}^{m+1,k} = (u_{N^-+1}^{m+1,k}, \ldots, u_{N^+-1}^{m+1,k})$$

denote the k-th iterate of the algorithm at the $(m+1)$-st time-step. (Thus, at time-step $m+1$, we start with the initial guess $\boldsymbol{u}^{m+1,0}$ and, as we apply the projected SOR algorithm, we generate $\boldsymbol{u}^{m+1,k+1}$ from $\boldsymbol{u}^{m+1,k}$. We know that $\boldsymbol{u}^{m+1,k} \to \boldsymbol{u}^{m+1}$ as $k \to \infty$.)

(i) Given $\boldsymbol{u}^m$, first form the vector $\boldsymbol{b}^m$ using formulæ (9.7) and (9.11), and calculate the constraint vector $\boldsymbol{g}^{m+1}$ using (9.4) and (9.10).

(ii) Start with the initial guess $\boldsymbol{u}^{m+1,0} = \max(\boldsymbol{u}^m, \boldsymbol{g}^{m+1})$, that is, $u_n^{m+1,0} = \max(u_n^m, g_n^{m+1})$.

(iii) In increasing n-indicial order, first form the quantity y_n^{k+1} by

$$y_n^{m+1,k+1} = \frac{1}{1+\alpha}\left(b_n^m + \tfrac{1}{2}\alpha(u_{n-1}^{m+1,k+1} + u_{n+1}^{m+1,k})\right)$$

(note, again, that we use the updated $u_{n-1}^{m+1,k+1}$ rather than the old $u_{n-1}^{m+1,k}$). Then generate $u_n^{m+1,k+1}$ using

$$u_n^{m+1,k+1} = \max\left(g_n^{m+1}, u_n^{m+1,k} + \omega(y_n^{m+1,k+1} - u_n^{m+1,k})\right),$$

where $1 < \omega < 2$ is the over-relaxation parameter.

(iv) Test whether or not $\|\boldsymbol{u}^{m+1,k+1} - \boldsymbol{u}^{m+1,k}\|$ is smaller than a pre-chosen tolerance ϵ, that is, test whether

$$\sum_n \left(u_n^{m+1,k+1} - u_n^{m+1,k}\right)^2 \le \epsilon^2.$$

If it is, go on to step (v). If it is not, go back to step (iii) and repeat the process with k replaced by $k+1$.

(v) When the vectors $\boldsymbol{u}^{m+1,k}$ have converged to the required tolerance, put $\boldsymbol{u}^{m+1} = \boldsymbol{u}^{m+1,k+1}$.

(vi) Return to step 1 until the necessary number of time-steps have been completed.

```
American( u,dt,dx,Nminus,Nplus,M,omega,eps )
{
  a = dt/(dx*dx);
  a2 = alpha/2.0;
  eps = 1.0e-8;
  omega = 1.0;
  domega = 0.05;
  oldloops = 10000;

  for( n=Nminus; n<=Nplus; ++n )
  {
    x[n] = n*dx;
    u[n] = pay_off( x[n], 0.0 );
  }

  for( m=1; m<=M; ++m )
  {
    tau = m*dt;

    for( n=Nminus+1; n<Nplus; ++n )
    {
      g[n] = payoff( x[n],tau );
      b[n] = u[n] + a2*(u[n+1]-2*u[n]+u[n-1]);
    }
    g[Nminus] = pay_off( x[Nminus],tau );
    g[ Nplus] = pay_off( x[ Nplus],tau );
    u[Nminus] = g[Nminus];
    u[ Nplus] = g[ Nplus];

    PSOR_solver( u,b,g,Nminus,Nplus,a,omega,eps,loops );
    if ( loops > oldloops ) domega *= -1.0;
    omega += domega;
    oldloops = loops;
  }
}
```

Figure 9.2 Pseudo-code for solution of an American option problem. Compare with Figures 8.11 and 8.14.

In Figure 9.2 we give a pseudo-code implementation of this algorithm.

Technical Point: Bermudan Options.

The projected SOR method is a generalisation of the SOR method. Indeed, the two methods are identical except that the step

$$u_n^{m+1,k+1} = u_n^{m+1,k} + \omega(y_n^{m+1,k+1} - u_n^{m+1,k})$$

that occurs in the SOR algorithm becomes

$$u_n^{m+1,k+1} = \max\left(g_n^{m+1}, u_n^{m+1,k} + \omega(y_n^{m+1,k+1} - u_n^{m+1,k})\right)$$

in the projected SOR algorithm.

One advantage of the SOR algorithm over the LU algorithm when valuing European options is that the computer code is identical to the code for an American version except for a single line. Thus, using SOR or projected SOR, it is trivial to modify the code for a European or American option to value an option which may be exercised early, but only on predetermined dates. Such an option is referred to as a **Bermudan option**. During a period where the option may be exercised early, the constraint that its value exceeds the payoff applies. This implies that we use projected SOR during such a period. When early exercise is forbidden, the payoff constraint does not apply, and we use ordinary SOR. The difference between the two algorithms amounts to using or not using the max function. For completeness, in Figure 9.3 we give a general projected SOR algorithm that allows the early exercise constraint to be turned either on or off at each time-step. This may be used to value Bermudan options.

9.6 Numerical Examples

In Figure 9.4 we give values for the American put with interest rate $r = 0.10$, volatility $\sigma = 0.4$ and exercise price $E = 10$. The calculation is carried out with $\alpha = 1$.

In Figure 9.5 we show a numerically computed solution of an American call problem with $E = 10$, $r = 0.25$, $\sigma = 0.8$, $D_0 = 0.2$ and lifetime of one year, together with the corresponding European value. The smooth separation of the option value from the payoff function can be seen at the point S_f. (We determine the position of the free boundary *a posteriori* by finding the x-node $x_n = n\,\delta x$ at which $u_n^m > g_n^m$ but for which $u_{n-1}^m \le g_{n-1}^m$. Further resolution is possible by assuming linear variations for $u(x, m\,\delta\tau)$ and $g(x, m\,\delta\tau)$ between grid points $n\,\delta x$; we can then approximate the position of the free boundary by the intersection of the straight line segments joining g_{n-1}^m, g_n^m and u_{n-1}^m, u_n^m.)

```
GSOR_solver( u,b,g,Nminus,Nplus,a,omega,eps,early,loops )
{
  loops = 0;
  a2 = a/2.0;
  u[Nminus] = g[Nminus];
  u[ Nplus] = g[ Nplus];

  do
  {
    error = 0.0;

    for( n=Nminus+1; n<Nplus; ++n )
    {
      y = (b[n]+a2*(u[n-1]+u[n+1]))/(1+a);

      if( early_exercise == TRUE)
        y = max( g[n], u[n]+omega*(y-u[n]) );
      else
        y = u[n]+omega*(y-u[n]);

      error += (u[n]-y)*(u[n]-y);
      u[n]=y;
    }
    ++loops;
  } while ( error>eps );
}
```

Figure 9.3 Pseudo-code for generalised SOR algorithm for American, European and Bermudan option problems using Crank–Nicolson finite-differences. Here `a` $= \alpha$, `a2` $= \frac{1}{2}\alpha$, `Nplus` $= N^+$, `Nminus` $= N^-$, `u[n]` $= u_n^{m+1,k}$ or $u_n^{m+1,k+1}$, `b[n]` $= b_n^m$, `g[n]` $= g_n^{m+1}$, $y = y_n^{m+1,k+1}$, `omega` $= \omega$ and `eps` is the desired error tolerance. The parameter `early` determines whether or not early exercise is allowed. If early exercise is allowed, `early` should be set to `TRUE` by the calling routine; projected SOR is then used. If early exercise is not allowed, `early` should be set to `FALSE`; SOR is then used. As the routine is called for each time-step a Bermudan option may be valued by calling the routine with `early` = `TRUE` at time-steps when the option may be exercised early and `early` = `FALSE` at time-steps when it may not. The routine returns the number of iterations required in `loops` so the calling program can optimize the value of `omega`.

Asset Price	Payoff Value	3 months		6 months	
		Amer.	Euro.	Amer.	Euro.
0.00	10.0000	10.0000	9.7531	10.0000	9.5123
2.00	8.0000	8.0000	7.7531	8.0000	7.5123
4.00	6.0000	6.0000	5.7531	6.0000	5.5128
6.00	4.0000	4.0000	3.7569	4.0000	3.5583
8.00	2.0000	2.0200	1.9024	2.0951	1.9181
10.00	0.0000	0.6913	0.6694	0.9211	0.8703
12.00	0.0000	0.1711	0.1675	0.3622	0.3477
14.00	0.0000	0.0332	0.0326	0.1320	0.1279
16.00	0.0000	0.0055	0.0054	0.0460	0.0448

Figure 9.4 Crank–Nicolson solution for an American put with $E = 10$, $r = 0.1$, $\sigma = 0.4$ and with expiry times of three and six months.

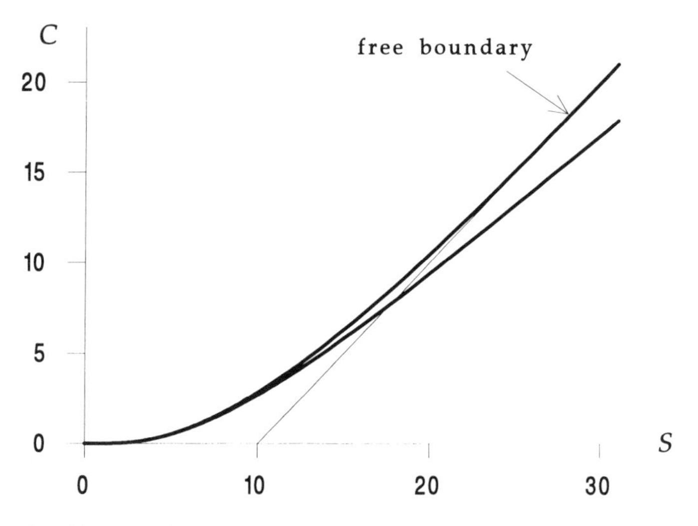

Figure 9.5 Numerically calculated solution of an American call problem with $E = 10$, $r = 0.25$, $D_0 = 0.2$, $\sigma = 0.8$ and expiry date of one year. The parameter values have been chosen to exaggerate the difference between the American option and its European counterpart (also shown).

9.7 Convergence of the Method

A detailed analysis of the convergence of the finite-difference approximation to the linear complementarity form of the American option problem is somewhat beyond the aims of this book. A rigorous proof of the convergence involves the use of a good deal of abstract functional analysis and is most easily presented within the framework of the finite-element solution of the variational inequality formulation of the problem. (The variational inequality formulation can be derived from, and is equivalent to, the linear complementarity formulation; see *Option Pricing* for more details.) As the numerical algorithms for solving the linear complementarity formulation by finite differences and for solving the variational inequality formulation by finite elements are identical the convergence of the algorithm presented here can be established in this manner. A detailed analysis of the convergence may be found in the literature.

Further Reading

- For the theoretical discussion (including the theoretical numerical analysis) of free boundary problems see Elliott & Ockendon (1982) and Kinderlehrer & Stampacchia (1980).
- For proofs of convergence of the projected SOR algorithm and of the finite-difference algorithm given in this chapter, see Elliott & Ockendon (1982).
- For the numerical solution of implicit free boundary problems by boundary tracking methods and for a general practical guide to the numerical solution of linear complementarity problems, see Crank (1984).
- The projected SOR algorithm was originally devised by Cryer (1971) in the context of quadratic optimization.
- The solution of American options using the explicit finite-difference method was first done by Brennan & Schwartz (1978).

Exercises

1. Write a computer program to implement the projected SOR method of valuing American options.

2. Show that the explicit finite-difference discretisation of the linear complementarity problem (9.1) leads to the equations

$$y_n^{m+1} = (1-2\alpha)u_n^m + \alpha(u_{n-1}^m + u_{n+1}^m),$$
$$u_n^{m+1} = \max\left(y_n^{m+1}, g_n^{m+1}\right).$$

Write a computer program to implement this scheme.

3. Show that the fully implicit finite-difference discretisation of (9.1) is

$$\mathbf{C}\boldsymbol{u}^m \geq \boldsymbol{b}^m, \quad \boldsymbol{u}^m \geq \boldsymbol{g}^m,$$
$$(\boldsymbol{u}^m - \boldsymbol{g}^m) \cdot (\mathbf{C}\boldsymbol{u}^m - \boldsymbol{b}^m) = 0.$$

where $\mathbf{C}$ is the tridiagonal matrix and $\boldsymbol{b}^m$ is the vector given, respectively, by

$$\begin{pmatrix} 1+2\alpha & -\alpha & & 0 \\ -\alpha & \ddots & \ddots & \\ & \ddots & \ddots & -\alpha \\ 0 & & -\alpha & 1+2\alpha \end{pmatrix}, \quad \begin{pmatrix} u_{N^-+1}^{m-1} \\ \vdots \\ u_0^{m-1} \\ \vdots \\ u_{N^+-1}^{m-1} \end{pmatrix} + \alpha \begin{pmatrix} g_{N^-}^m \\ 0 \\ \vdots \\ 0 \\ g_{N^+}^m \end{pmatrix}.$$

4. Describe the projected SOR method for the problem obtained in Exercise 3.

5. Use an explicit finite-difference discretisation of the unmodified linear complementarity problem for an American put option to obtain the finite-difference equations

$$W_n^m = a_n V_{n-1}^{m+1} + b_n V_n^{m+1} + c_n V_{n+1}^{m+1}$$
$$V_n^m = \max(E - n\,\delta S, W_n^m)$$

where

$$a_n = \tfrac{1}{2}\left(\sigma^2 n^2 - (r - D_0)n\right)\delta t$$
$$b_n = 1 - \left(\sigma^2 n^2 + r\right)\delta t$$
$$c_n = \tfrac{1}{2}\left(\sigma^2 n^2 + (r - D_0)n\right)\delta t,$$

E is the exercise price and $V_n^m \approx V(n\,\delta S, m\,\delta t)$ is the option value. (See Exercise 9 of the previous chapter.) What changes would be necessary to value an American call or a cash-or-nothing option?

6. Use a fully implicit finite-difference discretisation of the untransformed linear complementarity problem for an American put to obtain the constrained matrix problem

$$\mathbf{M}\boldsymbol{V}^m \geq \boldsymbol{V}^{m+1}, \quad \boldsymbol{V}^m \geq \Lambda,$$
$$(\boldsymbol{V}^m - \Lambda) \cdot \left(\mathbf{M}\boldsymbol{V}^m - \boldsymbol{V}^{m+1}\right) = 0$$

where

$$\boldsymbol{V}^m = (V_0^m, \ldots, V_N^m),$$

$$\Lambda = (E, \max(E - \delta S, 0), \max(E - 2\,\delta S, 0), \ldots, \max(E - N\,\delta S, 0)),$$

$$\mathbf{M} = \begin{pmatrix} B_0 & C_0 & 0 & \cdots & 0 \\ A_1 & B_1 & C_1 & & \vdots \\ 0 & A_2 & \ddots & \ddots & 0 \\ \vdots & & \ddots & \ddots & C_{N-1} \\ 0 & \cdots & 0 & A_N & B_N \end{pmatrix}$$

and

$$\begin{aligned} A_n &= -\tfrac{1}{2}\left(\sigma^2 n^2 - (r - D_0)n\right)\delta t \\ B_n &= 1 + \left(\sigma^2 n^2 + r\right)\delta t \\ C_n &= -\tfrac{1}{2}\left(\sigma^2 n^2 + (r - D_0)n\right)\delta t. \end{aligned}$$

(See Exercises 13 and 15 of the previous chapter.)

7. Describe a projected SOR algorithm for the constrained matrix problem in Exercise 6.

8. Write a routine that will value European, American and Bermudan options, using a fully implicit finite-difference method.

10 Binomial Methods

10.1 Introduction

Binomial methods for valuing options and other derivative securities arise from discrete random walk models of the underlying security. They rely only indirectly on the Black–Scholes analysis through the assumption of risk neutrality; their relation to the partial differential equation and inequality models described and derived earlier in this book becomes evident only when it is seen that binomial methods are particular cases of the explicit finite-difference method described in Chapter 8 (see Exercise 5).

There are two main ideas underlying the binomial methods. The first of these is that the continuous random walk (2.1) may be modelled by a discrete random walk with the following properties:

- The asset price S changes only at the discrete times δt, $2\,\delta t$, $3\,\delta t, \ldots$, up to $M\delta t = T$, the expiry date of the derivative security. We use δt instead of dt to denote the small but *non-infinitesimal* time-step between movements in the asset price.
- If the asset price is S^m at time $m\,\delta t$ then at time $(m+1)\,\delta t$ it may take one of only two possible values, $\mathfrak{u}\,S^m > S^m$ or $\mathfrak{d}\,S^m < S^m$. (That is, during a single time-step, the asset price may move from S up to $\mathfrak{u}\,S$ or down to $\mathfrak{d}\,S$; see Figure 10.1). Note that this is equivalent to assuming that there are only two returns $\delta S/S$ possible at each time-step, $\mathfrak{u}-1>0$ and $\mathfrak{d}-1<0$, and that these two returns are the same for all time-steps.
- The probability, $\mathfrak{p}$, of S moving up to $\mathfrak{u}S$ is known (as is the probability $(1-\mathfrak{p})$ of S moving down to $\mathfrak{d}S$).

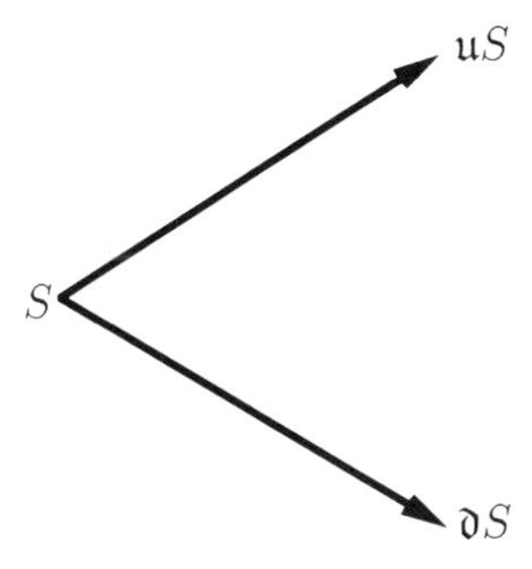

Figure 10.1. Asset price movement in the binomial method.

As we shall see, we choose the parameters[1] $\mathfrak{u}$, $\mathfrak{d}$ and $\mathfrak{p}$ in such a way that the important statistical properties of the discrete random walk described above coincide with those of a (slightly modified version of) the continuous random walk (2.1).

Starting with a given value of the asset price (for example, today's asset price) the remaining life-time of the derivative security is divided up into M time-steps of size $\delta t = (T - t)/M$. The asset price S is assumed to move only at times $m\,\delta t$ for $m = 1, 2, \ldots, M$. Then a **tree** of all possible asset prices is created. This tree is constructed by starting with the given value S, generating the two possible asset prices ($\mathfrak{u}S$ and $\mathfrak{d}S$) at the first time-step, then the three possible asset prices ($\mathfrak{u}^2 S$, $\mathfrak{u}\mathfrak{d}S = \mathfrak{d}\mathfrak{u}S$ and $\mathfrak{d}^2 S$) at the second time-step, and so forth until the expiry time of the security is reached; see Figure 10.2. Note that as an up-jump followed by a down-jump leads to the same asset value as a down-jump followed by an up-jump the binomial tree reconnects and after m time-steps there are only $m + 1$ possible asset prices.

The second assumption is that of a risk-neutral world, that is, one where an investor's risk preferences are irrelevant to derivative security valuation. This assumption may be made whenever it is possible to hedge a portfolio perfectly and make it riskless (see Section 5.6). Under these circumstances we may assume that investors are risk-neutral (even though most are not), and that the return from the underlying is the risk-free interest rate. The μ in the stochastic differential equation $dS = \sigma S\,dX + \mu S\,dt$ is a measure of the expected growth rate of the underlying asset and, as we have seen, it does not enter into the Black–Scholes

[1] Although we assume that $\mathfrak{u}$, $\mathfrak{d}$ and $\mathfrak{p}$ are independent of time t and asset price S, this is not an essential feature of the binomial method.

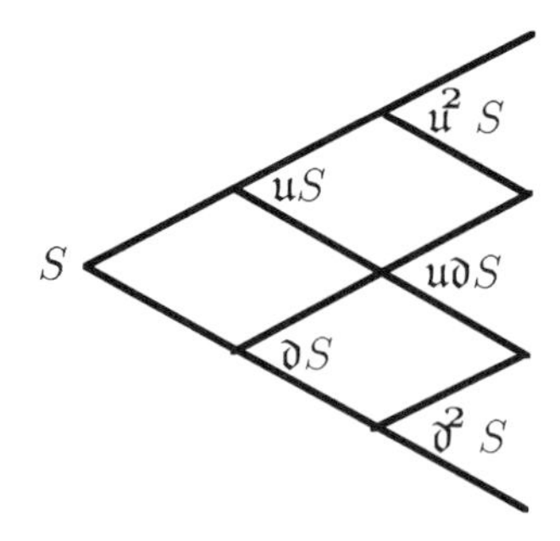

Figure 10.2. The binomial tree of possible asset prices.

equation. Recalling the discussion of Section 5.6, in a risk-neutral world we replace the stochastic differential equation (2.1) by

$$\frac{dS}{S} = \sigma\, dX + r\, dt. \tag{10.1}$$

We choose values for $\mathfrak{u}$, $\mathfrak{d}$ and $\mathfrak{p}$ in our discrete random walk to reflect the important statistical properties of the continuous random walk (10.1), i.e. with a growth rate r instead of μ.

Under the assumption of a risk-neutral world we observe that the value, V^m, of the derivative security at time-step $m\,\delta t$ is the *expected value* of the security at time-step $(m+1)\,\delta t$ discounted by the risk-free interest rate r:

$$V^m = \mathcal{E}\left[e^{-r\delta t}V^{m+1}\right]; \tag{10.2}$$

this is another way of interpreting the Black–Scholes formula (5.17).

In a binomial method, we first build a tree of possible values of asset prices and their probabilities, given an initial asset price, then use this tree to determine the possible asset prices at expiry and the probabilities of these asset prices being realised. The possible values of the security at expiry can then be calculated,[2] and, by working back down the tree using (10.2), the security can be valued. A useful consequence is that we can quite easily deal with the possibility of early exercise and with dividend payments.

[2] We are assuming here that the payoff function is determined only by the value of the underlying at expiry. This is not the case, for example, for a path-dependent option such as a lookback or an average strike.

10.2 The Discrete Random Walk

The probability, $\mathfrak{p}$, of an up-jump and the jump sizes $\mathfrak{u}$ and $\mathfrak{d}$ are chosen so that the discrete random walk represented by the tree and the continuous random walk (10.1) have the same mean and variance. That is, given that the value of the asset is S^m at time-step $m\,\delta t$, we equate the expected values and variances of S^{m+1} (where S^{m+1} is the asset value at time-step $(m+1)\delta t$) under the continuous risk-neutral random walk (10.1) and the discrete binomial model.

Given that the asset value is S^m at time-step $m\,\delta t$, the expected value of S^{m+1} under the continuous random walk model (10.1) is (see Exercise 1)

$$\mathcal{E}_c[S^{m+1}|S^m] = \int_0^\infty S' p\Big(S^m, m\,\delta t; S', (m+1)\,\delta t\Big)\, dS' = e^{r\,\delta t} S^m$$

where $p(S,t;S',t')$ is the probability density function

$$p(S,t;S',t') = \frac{1}{\sigma S'\sqrt{2\pi(t'-t)}} e^{-\left(\log(S'/S)-(r-\frac{1}{2}\sigma^2)(t'-t)\right)^2/2\sigma^2(t'-t)} \tag{10.3}$$

for the risk-neutral random walk (10.1) (see Section 2.3). The expected value of S^{m+1}, given S^m, under the discrete binomial random walk is

$$\mathcal{E}_b[S^{m+1}|S^m] = \Big(\mathfrak{p}\mathfrak{u} + (1-\mathfrak{p})\mathfrak{d}\Big) S^m.$$

Equating these two expected values gives

$$\mathfrak{p}\mathfrak{u} + (1-\mathfrak{p})\mathfrak{d} = e^{r\,\delta t}. \tag{10.4}$$

The variance of S^{m+1}, given S^m, is defined to be

$$\mathrm{var}[S^{m+1}|S^m] = \mathcal{E}[(S^{m+1})^2|S^m] - \mathcal{E}[S^{m+1}|S^m]^2.$$

Under the continuous random walk (10.1) we have (see Exercise 1)

$$\begin{aligned}\mathcal{E}_c[(S^{m+1})^2|S^m] &= \int_0^\infty (S')^2 p\Big(S^m, m\,\delta t; S', (m+1)\,\delta t\Big)\, dS' \\ &= e^{(2r+\sigma^2)\delta t}(S^m)^2,\end{aligned}$$

where $p(S,t;S',t')$ is the density function (10.3). Thus the variance under the continuous process (10.1) is

$$\mathrm{var}_c[S^{m+1}|S^m] = e^{2r\,\delta t}(e^{\sigma^2\delta t} - 1)(S^m)^2.$$

Under the discrete binomial process we have

$$\mathcal{E}_b[(S^{m+1})^2|S^m] = \Big(\mathfrak{p}\mathfrak{u}^2 + (1-\mathfrak{p})\mathfrak{d}^2\Big)(S^m)^2$$

and, using (10.4), $\mathcal{E}_b[S^{m+1}|S^m] = S^m e^{r\,\delta t}$. We therefore have

$$\mathrm{var}_b[S^{m+1}|S^m] = \left(\mathfrak{p}\mathfrak{u}^2 + (1-\mathfrak{p})\mathfrak{d}^2 - e^{2r\,\delta t}\right)(S^m)^2.$$

Equating these two variances we find that

$$\mathfrak{p}\mathfrak{u}^2 + (1-\mathfrak{p})\mathfrak{d}^2 = e^{(2r+\sigma^2)\delta t}. \qquad (10.5)$$

Equations (10.4) and (10.5) are two equations for the three unknowns $\mathfrak{u}$, $\mathfrak{d}$ and $\mathfrak{p}$. In order to determine these three unknowns uniquely we require another equation. As (10.4) and (10.5) determine all the statistically important properties of the discrete random walk (except, possibly, the requirements that $\mathfrak{u} > 0$, $\mathfrak{d} > 0$ and $0 \le \mathfrak{p} \le 1$), the choice of the third equation is somewhat arbitrary. There are two popular choices,

$$\mathfrak{u} = \frac{1}{\mathfrak{d}}, \qquad (10.6)$$

and

$$\mathfrak{p} = \tfrac{1}{2}. \qquad (10.7)$$

10.2.1 The Case $\mathfrak{u} = 1/\mathfrak{d}$

In this case we obtain a binomial method where $\mathfrak{u}$, $\mathfrak{d}$ and $\mathfrak{p}$ are determined by (10.4), (10.5) and (10.6). From (10.4) and (10.5) we find that

$$\mathfrak{p} = \frac{e^{r\,\delta t} - \mathfrak{d}}{\mathfrak{u} - \mathfrak{d}} = \frac{e^{(2r+\sigma^2)\delta t} - \mathfrak{d}^2}{\mathfrak{u}^2 - \mathfrak{d}^2}, \qquad (10.8)$$

so that

$$\mathfrak{u} + \mathfrak{d} = \frac{e^{(2r+\sigma^2)\delta t} - \mathfrak{d}^2}{e^{r\,\delta t} - \mathfrak{d}}.$$

Using (10.6) to eliminate $\mathfrak{u}$ then gives the quadratic equation[3]

$$\mathfrak{d}^2 - 2A\mathfrak{d} + 1 = 0$$

where

$$A = \tfrac{1}{2}\left(e^{-r\,\delta t} + e^{(r+\sigma^2)\delta t}\right). \qquad (10.9)$$

Solving for $\mathfrak{d}$ and using (10.6) to determine $\mathfrak{u}$ and (10.8) to find $\mathfrak{p}$, we find

$$\mathfrak{d} = A - \sqrt{A^2-1}, \quad \mathfrak{u} = A + \sqrt{A^2-1}, \quad \mathfrak{p} = \frac{e^{r\,\delta t} - \mathfrak{d}}{\mathfrak{u} - \mathfrak{d}}. \qquad (10.10)$$

[3] In fact, it is easy to see that $\mathfrak{u}$ also satisfies the same quadratic equation, i.e., $\mathfrak{u}^2 - 2A\mathfrak{u} + 1 = 0$.

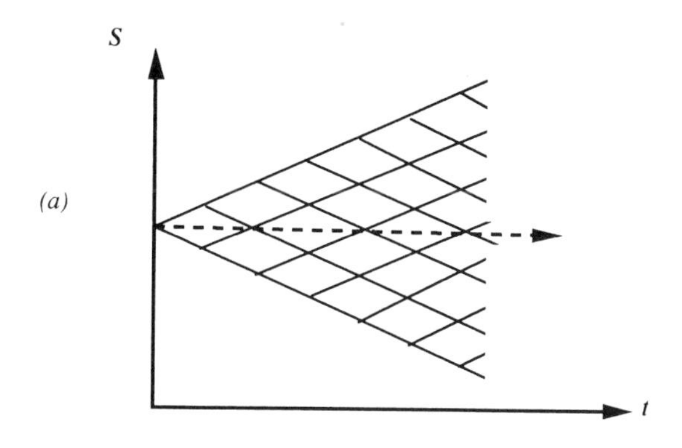

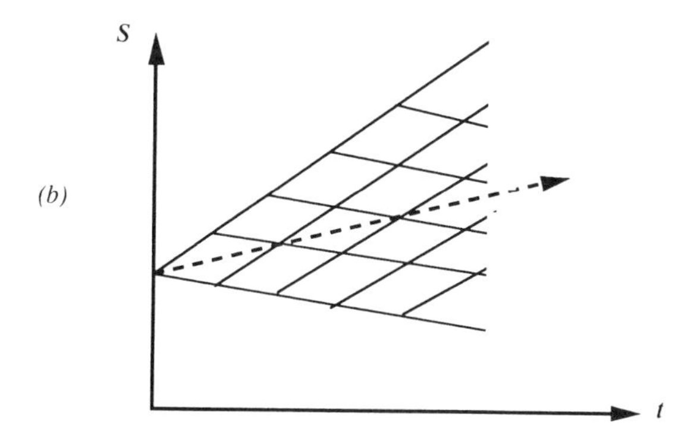

Figure 10.3. Binomial trees when (a) $\mathfrak{u} = 1/\mathfrak{d}$ and (b) $\mathfrak{p} = \frac{1}{2}$.

If too large a time-step is taken, $\mathfrak{p}$ or $1 - \mathfrak{p}$ may become negative, in which case the binomial method will fail.

The choice (10.6) leads to a tree in which the starting asset price re-occurs every even time-step and which is symmetric about this price; see Figure (10.3a). The asset price drift, caused by the $r\,dt$ term in (10.1), is reflected in the fact that the probability of an up-jump differs from the probability of a down-jump; $\mathfrak{p} \neq 1 - \mathfrak{p}$.

10.2.2 The Case $\mathfrak{p} = \frac{1}{2}$

In this case, the constants $\mathfrak{u}$ are $\mathfrak{d}$ determined by (10.4), (10.5) and $\mathfrak{p}$ is given by (10.7). Thus we find that

$$\mathfrak{u} + \mathfrak{d} = 2e^{r\,\delta t}, \quad \mathfrak{u}^2 + \mathfrak{d}^2 = 2e^{(2r+\sigma^2)\delta t}. \qquad (10.11)$$

The equations are clearly invariant under interchange of $\mathfrak{u}$ and $\mathfrak{d}$, so we look for solutions of the form $\mathfrak{u} = B + C$, $\mathfrak{d} = B - C$, to find that

$$\begin{aligned} \mathfrak{d} &= e^{r\,\delta t}\left(1 - \sqrt{e^{\sigma^2\,\delta t} - 1}\right), \\ \mathfrak{u} &= e^{r\,\delta t}\left(1 + \sqrt{e^{\sigma^2\,\delta t} - 1}\right), \\ \mathfrak{p} &= \tfrac{1}{2}. \end{aligned} \tag{10.12}$$

The probabilities of an up-jump and a down-jump are equal in this case and we find that $\mathfrak{u}\mathfrak{d} > 1$ (assuming $r > 0$ and δt is not too large) and the tree is oriented in the direction of the drift (see Figure 10.3b). If too large a time-step is taken $\mathfrak{d}$ may become negative, in which case the binomial method will fail.

10.2.3 The Binomial Tree

Using either (10.10) or (10.12), we may build up a tree of possible asset prices. We start at the current time $t = 0$. We assume that at this time we know the asset price, S_0^0. Then at the next time-step δt there are two possible asset prices, $S_1^1 = \mathfrak{u}S_0^0$ and $S_0^1 = \mathfrak{d}S_0^0$. At the following time-step, $2\,\delta t$, there are three possible asset prices $S_2^2 = \mathfrak{u}^2 S_0^0$, $S_1^2 = \mathfrak{u}\mathfrak{d}S_0^0 = \mathfrak{d}\mathfrak{u}S_0^0$ and $S_0^2 = \mathfrak{d}^2 S_0^0$. At the third time-step $3\,\delta t$ there are four possible asset prices, and so forth. At the m-th time-step $m\,\delta t$ there are $m + 1$ possible values of the asset price,

$$S_n^m = \mathfrak{d}^{m-n}\mathfrak{u}^n S_0^0, \quad n = 0, 1, \ldots, m.$$

(Note that here S_n^m denotes the n-th possible value of S at time-step $m\,\delta t$ whereas $\mathfrak{d}^n$ and $\mathfrak{u}^n$ denote $\mathfrak{d}$ and $\mathfrak{u}$ raised to the n-th power.) At the final time-step, $M\,\delta t$, we have $M + 1$ possible values of the underlying asset. In the case $\mathfrak{u} = 1/\mathfrak{d}$ we see that

$$S_n^m = \mathfrak{u}^{2n-m} S_0^0 = \mathfrak{d}^{m-2n} S_0^0, \quad n = 0, 1, 2, \ldots, m$$

so that whenever m is even, $S_{m/2}^m = S_0^0$.

Note that the trees in Figure 10.3 reconnect. This has two consequences that are of immediate interest. The first is that the history of a particular asset price is lost, as there is clearly more than one path to any given point. Thus, in general, path-dependent options cannot be valued using these reconnecting trees. The second is that the total number of lattice points increases only quadratically with the number of time-steps. This means that a large number of time-steps can be taken.

10.3 Valuing the Option

Assuming that we know the payoff function for our derivative security, and that it depends only on the values of the underlying asset at expiry, we are able to value it at expiry, i.e. time-step $M\,\delta t$. If we are considering a put, for example, we find that

$$V_n^M = \max(E - S_n^M, 0), \quad n = 0, 1, \ldots, M, \tag{10.13}$$

where E is the exercise price and V_n^M denotes the n-th possible value of the put at time-step M and the n-th possible asset value S_n^M. For a call, we find that

$$V_n^M = \max(S_n^M - E, 0), \quad n = 0, 1, \ldots, M, \tag{10.14}$$

and, similarly, for a cash or nothing call, with exercise price E and payoff

$$V(S,T) = \begin{cases} 0 & S < E, \\ B & S \geq E, \end{cases}$$

we have

$$V_n^M = \begin{cases} 0 & S_n^M < E, \\ B & S_n^M \geq E, \end{cases} \qquad n = 0, 1, \ldots, M. \tag{10.15}$$

We can find the expected value of the derivative security at the time-step prior to expiry, $(M-1)\delta t$, and for possible asset price S_n^{M-1}, $n = 0, 1, \ldots M-1$, since we know that the probability of an asset priced at S_n^{M-1} moving to S_{n+1}^M during a time-step is $\mathfrak{p}$, and the probability of it moving to S_n^M is $(1-\mathfrak{p})$. Using the risk-neutral argument we can calculate the value of the security at each possible asset price for time-step $(M-1)$. Similarly this allows us to find the value of the security at time-step $(M-2)$, and so on, back to time-step 0. This gives us the value of our security at the current time.

10.4 European Options

Let V_n^m denote the value of the option at time-step $m\,\delta t$ and asset price S_n^m (where $0 \leq n \leq m$). We calculate the expected value of the option at time-step $m\,\delta t$ from the values at time-step $(m+1)\delta t$ and discount this to obtain the present value using the risk-free interest rate r,

$$e^{r\delta t} V_n^m = \mathfrak{p} V_{n+1}^{m+1} + (1-\mathfrak{p}) V_n^{m+1}.$$

```
euro_option( array,S0,u,d,p,r,dt,M )
{
  discount = exp(-r*dt);

  array[0] = S0;
  for( m=1; m<=M; ++m )
  {
    for( n=m; n>0; --n )
      array[n] = u*array[n-1];
    array[0] = d*array[0];
  }

  for( n=0; n<=M; ++n )
    array[n] = pay_off( array[n] );

  for( m=M; m>0; --m )
  {
    for( n=0; n<m; ++n )
    {
      tmp = p*array[n+1] + (1-p)*array[n];
      array[n] = discount*tmp;
    }
  }
}
```

Figure 10.4 Pseudo-code for binomial model for a simple European option. `array[]` is an array used to store the asset prices during the construction of the tree (the first `for` loop) and the option values (the final `for` loop). `S0` is the current value of the underlying, `u` $= \mathfrak{u}$, `d` $= \mathfrak{d}$ and `p` $= \mathfrak{p}$; these latter three may be calculated using either (10.9) and (10.10) from Section 10.2.1 or (10.12) from Section 10.2.2. `r` is the interest rate, `dt` is the time-step and `M` the number of time-steps. The routine first builds a tree of possible asset prices, then finds the values of the option at expiry using the `pay_off` function. Finally it calculates the present values of the expected values of the option price, under a risk-neutral random walk, from expiry back until the present. The present value of the option is returned in `array[0]`.

This gives

$$V_n^m = e^{-r\delta t}\left(\mathfrak{p}V_{n+1}^{m+1} + (1-\mathfrak{p})V_n^{m+1}\right), \qquad n = 0, 1, \ldots, m. \qquad (10.16)$$

As we know the value of V_n^M, $n = 0, 1, \ldots, M$ from the payoff function we can recursively determine the values of V_n^m for each $n = 0, 1, \ldots, m$ for $m < M$ to arrive at the current value of the option, V_0^0.

We do not require the asset prices S_n^m during the evaluation of European option prices, other than the S_n^M when finding V_n^M. At each time-step we can discard the old S_n^m as soon as we have calculated the S_n^{m+1}. Once the V_n^M have been found, we can discard the S_n^M as well.

T	Black–Scholes	Binomial Method ($\mathfrak{u} = 1/\mathfrak{d}$) $M = 16$	32	64	128	256
0.25	4.8511	4.8511	4.8511	4.8511	4.8511	4.8511
0.50	4.7048	4.7046	4.7047	4.7047	4.7048	4.7048
0.75	4.5636	4.5626	4.5632	4.5634	4.5634	4.5636
1.00	4.4304	4.4292	4.4300	4.4300	4.4300	4.4304

Figure 10.5 Comparison of binomial method (with $\mathfrak{u} = 1/\mathfrak{d}$) and Black–Scholes values for a European put with $E = 10$, $S = 5$, $r = 0.06$ and $\sigma = 0.3$. Expiry time T is measured in years.

T	Black–Scholes	Binomial Method ($\mathfrak{p} = \frac{1}{2}$) $M = 16$	32	64	128	256
0.25	4.8511	4.8511	4.8511	4.8511	4.8511	4.8511
0.50	4.7048	4.7046	4.7047	4.7047	4.7048	4.7048
0.75	4.5636	4.5625	4.5632	4.5634	4.5635	4.5635
1.00	4.4304	4.4293	4.4300	4.4300	4.4302	4.4303

Figure 10.6 Comparison of binomial method (with $\mathfrak{p} = \frac{1}{2}$) and Black–Scholes values for a European put with $E = 10$, $S = 5$, $r = 0.12$ and $\sigma = 0.5$. The time to expiry, T, is measured in years.

This observation leads to an extremely memory-efficient algorithm; the memory required varies linearly with the number of time-steps and the execution time varies quadratically with the number of time-steps. A pseudo-code for the algorithm is given in Figure 10.4.

In Figure 10.5 we compare the values of a European put, calculated using the binomial method with $\mathfrak{u} = 1/\mathfrak{d}$ (see Section 10.2.1), using $M = 16$, 32, 64, 128 and 256 time-steps, with the Black–Scholes value. In Figure 10.6 we compare the values of a European put, calculated using the binomial method with $\mathfrak{p} = \frac{1}{2}$ (see Section 10.2.2), using $M = 16$, 32, 64, 128 and 256 time-steps, with the Black–Scholes value.

10.5 American Options

We can easily incorporate the possibility of early exercise of an option into the binomial model. As before, we divide the time to expiry into M

```
amer_option( s,v,S0,u,d,p,r,dt,M )
{

  discount = exp(-r*dt);

  s[0][0] = S0;
  for( m=1; m<=M; ++m )
  {
    for( n=m+1; n>0; --n )
      s[m][n] = u*s[m-1][n-1];
    s[m][0] = d*s[m-1][0];
  }

  for( n=0; n<=M; ++n )
    v[M][n] = pay_off(s[M][n]);

  for( m=M; m>0; --m )
  {
    for( n=0; n<=m; ++n )
    {
      hold =  (1-p)*v[m+1][n] + p*v[m+1][n+1];
      hold *= discount;
      v[m][n] = max( hold,pay_off( s[m][n] ) );
    }
  }
}
```

Figure 10.7 Pseudo-code for binomial model for an American option. The arrays `s[][]` and `v[][]` are used to store the trees of asset values and option values, respectively. `S0` is the current value of the asset, `u` $= \mathfrak{u}$, `d` $= \mathfrak{d}$ and `p` $= \mathfrak{p}$; these latter three variables may be calculated using (10.10) and (10.11) (see Section 10.2.1) or (10.12) (see Section 10.2.2). The interest rate is `r`, the time-step `dt` and the number of time-steps `M`. The routine first builds and stores the tree of asset prices, then calculates the payoff at expiry using `pay_off` and finally values the option by taking the maximum of the expected value and the payoff for early exercise at each time-step and asset price.

equal time-steps $\delta t = T/M$ and build our tree of possible stock prices,

$$S_n^m, \quad n = 0, 1, \ldots, m,$$

where $S_{0,0}$ is the current value and S_n^m is a possible value at time-step $m\,\delta t$. As in the previous sections we may calculate $\mathfrak{u}$, $\mathfrak{d}$ and $\mathfrak{p}$ either using (10.9) and (10.10) or using (10.12). At time $M\,\delta t$ we can calculate the possible values of the option from the payoff function, for example, for puts using (10.13), for calls using (10.14) and for cash-or-nothing calls using (10.15).

Consider the situation at time-step m and at asset price S_n^m.

The option can be exercised prior to expiry to yield a profit determined by the payoff function; for puts, calls and cash-or-nothing calls, respectively, these are

$$\max(E - S_n^m, 0), \quad \max(S_n^m - E, 0), \quad \begin{cases} 0 & S_n^m < E \\ B & S_n^m \geq E \end{cases}.$$

If the option is retained, its value V_n^m is, as in the European case,

$$V_n^m = e^{-r\delta t}\left(\mathfrak{p}V_{n+1}^{m+1} + (1-\mathfrak{p})V_n^{m+1}\right).$$

The value of the option is the maximum of these two possibilities, i.e.,

$$V_n^m = \max\Big(\text{pay\_off}\left(S_n^m\right), e^{-r\,\delta t}\left(\mathfrak{p}V_{n+1}^{m+1} + (1-\mathfrak{p})V_n^{m+1}\right)\Big). \tag{10.17}$$

For example, for a put we have

$$V_n^m = \max\Big(\max\left(E - S_n^m, 0\right), e^{-r\,\delta t}\left(\mathfrak{p}V_{n+1}^{m+1} + (1-\mathfrak{p})V_n^{m+1}\right)\Big)$$

and for a call

$$V_n^m = \max\Big(\max\left(S_n^m - E, 0\right), e^{-r\,\delta t}\left(\mathfrak{p}V_{n+1}^{m+1} + (1-\mathfrak{p})V_n^{m+1}\right)\Big).$$

Implementing the scheme is almost as simple as the European case. A tree of asset values, S_n^m, is built first and, unlike the European case, saved. We then evaluate V_n^M from the payoff function, and work back down the tree to find the value of the option. The only additional complication is that it is necessary to test to decide which of the two possible values (early exercise or retaining the option) is greater. This is the reason for storing the S_n^m values as it allows us to implement this test efficiently. A pseudo-code that implements this algorithm is shown in Figure 10.7. The need to store the S_n^m implies that the memory requirements vary quadratically with the number of time-steps, as does the execution time.

In Figures 10.8 and 10.9 we give binomial approximations to an American put with exercise price 10, current asset price 9, interest rate of $r = 0.06$, volatility of $\sigma = 0.3$, for $M = 16$, $M = 32$, $M = 64$, $M = 128$ and $M = 256$ time-steps. In Figure 10.8 we have used (10.9) and (10.10) to calculate $\mathfrak{u}$, $\mathfrak{d}$ and $\mathfrak{p}$, and in Figure 10.9 we have used (10.12).

10.6 Dividend Yields

The binomial method can easily accommodate a constant dividend yield D_0 paid on the underlying. The effective risk-free growth rate of the

	Binomial Method ($\mathfrak{u} = 1/\mathfrak{d}$)				
T	$M = 16$	32	64	128	256
0.25	1.1316	1.1240	1.1261	1.1253	1.1260
0.50	1.2509	1.2598	1.2539	1.2553	1.2546
0.75	1.3579	1.3568	1.3569	1.3555	1.3547
1.00	1.4444	1.4332	1.4384	1.4342	1.4349

Figure 10.8 Comparison of binomial values (with $\mathfrak{u} = 1/\mathfrak{d}$) for an American put with $E = 10$, $S = 9$, $r = 0.06$ and $\sigma = 0.3$. Time to expiry, T, is measured in years.

	Binomial Method ($\mathfrak{p} = \frac{1}{2}$)				
T	$M = 16$	32	64	128	256
0.25	1.1311	1.1249	1.1266	1.1258	1.1260
0.50	1.2526	1.2590	1.2553	1.2559	1.2553
0.75	1.3609	1.3530	1.3573	1.3540	1.3541
1.00	1.4473	1.4358	1.4369	1.4358	1.4354

Figure 10.9 Comparison of binomial values (with $\mathfrak{p} = \frac{1}{2}$) for an American put with $E = 10$, $S = 9$, $r = 0.06$ and $\sigma = 0.3$. Time to expiry, T, is measured in years.

asset becomes $r - D_0$ rather than r, that is,

$$\frac{dS}{S} = (r - D_0)\,dt + \sigma\,dX.$$

Therefore we replace r by $r - D_0$ in the tree construction phase, when we calculate $\mathfrak{u}$, $\mathfrak{d}$ and $\mathfrak{p}$. In this way, (10.9) and (10.10), for the case $\mathfrak{u} = 1/\mathfrak{d}$, become

$$A = \tfrac{1}{2}\left(e^{-(r-D_0)\delta t} + e^{(r-D_0+\sigma^2)\delta t}\right) \tag{10.18}$$

and

$$\mathfrak{d} = A - \sqrt{A^2 - 1}, \quad \mathfrak{u} = A + \sqrt{A^2 - 1}, \quad \mathfrak{p} = \frac{e^{(r-D_0)\delta t} - \mathfrak{d}}{\mathfrak{u} - \mathfrak{d}}, \tag{10.19}$$

T	Black–Scholes	Binomial Method ($\mathfrak{u} = 1/\mathfrak{d}$)				
		$M = 16$	32	64	128	256
0.25	2.1116	2.1138	2.1128	2.1111	2.1113	2.1117
0.50	2.2820	2.2874	2.2856	2.2839	2.2818	2.2819
0.75	2.4374	2.4479	2.4316	2.4381	2.4361	2.4380
1.00	2.5752	2.5795	2.5806	2.5789	2.5771	2.5752

Figure 10.10 Comparison of binomial method (with $\mathfrak{u} = 1/\mathfrak{d}$) and Black–Scholes values for a European call with $E = 10$, $S = 12$, $r = 0.06$, $D_0 = 0.04$ and $\sigma = 0.3$. Expiry time T is measured in years.

and (10.12), for the case $\mathfrak{p} = \frac{1}{2}$, becomes

$$\begin{aligned}\mathfrak{d} &= e^{(r-D_0)\delta t}\left(1 - \sqrt{e^{\sigma^2\,\delta t} - 1}\right),\\ \mathfrak{u} &= e^{(r-D_0)\delta t}\left(1 + \sqrt{e^{\sigma^2\,\delta t} - 1}\right),\\ \mathfrak{p} &= \tfrac{1}{2}.\end{aligned} \qquad (10.20)$$

The present value of an asset is still determined by discounting using the risk-free interest rate r, so that (10.16) for European options and (10.17) for American options remain valid when calculating the expected present value of the option. Thus the only effect of continuous dividend yields is to modify the probability $\mathfrak{p}$ and jump sizes $\mathfrak{u}$ and $\mathfrak{d}$. Therefore, we can use the same codes as given above to value such options by simply modifying the parameters `u`, `d` and `p` according to (10.18), (10.19) and (10.20). This applies to both European and American style options.

In Figures 10.10 and 10.11 we compare exact Black–Scholes values and binomial approximations to a European call with exercise price 10, current asset price 12, interest rate of $r = 0.06$, dividend yield $D_0 = 0.03$ and volatility of $\sigma = 0.3$, for $M = 16$, $M = 32$, $M = 64$, $M = 128$ and $M = 256$ time-steps. In Figure 10.10 we have used (10.18) and (10.19) to calculate $\mathfrak{u}$, $\mathfrak{d}$ and $\mathfrak{p}$, and in Figure 10.11 we have used (10.20).

Further Reading

- For details of the binomial method as a model of asset prices see Cox & Rubinstein (1985).
- Hull (1993) and Wilmott *et al.* (1993) discuss the binomial valuation of options on assets paying discrete dividends.

T	Black–Scholes	Binomial Method ($\mathfrak{p} = \frac{1}{2}$) $M = 16$	32	64	128	256
0.25	2.1116	2.1132	2.1123	2.1121	2.1119	2.1118
0.50	2.2820	2.2906	2.2839	2.2824	2.2831	2.2825
0.75	2.4374	2.4436	2.4406	2.4407	2.4390	2.4368
1.00	2.5752	2.5667	2.5840	2.5745	2.5741	2.5756

Figure 10.11 Comparison of binomial method (with $\mathfrak{p} = \frac{1}{2}$) and Black–Scholes values for a European call with $E = 10$, $S = 12$, $r = 0.12$, $D_0 = 0.04$ and $\sigma = 0.5$. The time to expiry, T, is measured in years.

Exercises

1. Show that
$$\int_0^\infty S' p\Big(S^m, m\,\delta t; S', (m+1)\,\delta t\Big)\, dS' = S^m e^{r\,\delta t},$$
$$\int_0^\infty (S')^2 p\Big(S^m, m\,\delta t; S', (m+1)\,\delta t\Big)\, dS' = (S^m)^2 e^{(2r+\sigma^2)\delta t},$$
where $p(S, t; S', t')$ is given by (10.3). (Hint: put $S = e^x$ and then complete the square in the exponent.)

2. Show that the memory required to value a European option by the binomial model varies linearly with the number of time-steps, and that the execution time varies quadratically with the number of time-steps.

3. Write a computer program to value a European option with an arbitrary payoff using a binomial method. Assume that the underlying asset pays a continuous dividend yield.

4. Write a computer program to value an American option with an arbitrary payoff using a binomial method. Assume that the underlying asset pays a continuous dividend yield.

5. Show that the binomial method for European options, using (10.9) and (10.10), can be interpreted as an explicit finite-difference method for the Black–Scholes equation. (Hint: let $x_n^m = \log S_n^m$.)

6. Consider an option on a share that pays a discrete dividend yield of $d_y S(t_d)$ at time t_d. When the share goes ex-dividend, its value falls from S at t_d^- immediately before the dividend payment to $(1 - d_y)S$ at t_d^+ immediately after. How could you modify the binomial method to deal with such an option? (Hint: the dividend yield affects only the tree construction phase of the binomial method.)

Part three

Further Option Theory

11 Exotic and Path-dependent Options

11.1 Introduction

If the simple derivatives that we have so far considered, which are almost all variations on vanilla European or American calls and puts, were the only derivative securities that needed to be modelled, this book would not be worth writing, nor would the subject have much mathematical interest. This part of the book is devoted to an introductory *tour d'horizon* of some of the huge variety of seemingly complex derivative instruments that have been created and are traded; each poses a challenge both to the mathematical modeller and to the people who trade and hedge them in practice.

In this chapter we give an overview of some of the more common exotic and path-dependent options. In subsequent chapters we consider partial differential equation models for their valuation. In Chapter 13, in particular, we introduce a very simple but general framework for valuing many different path-dependent options. With this goal in mind, it is useful here to consider a classification of the varieties of exotic and path-dependent options.

A **path-dependent option** is an option whose payoff at exercise or expiry depends, in some non-trivial way, on the past history of the underlying asset price as well as its spot price at exercise or expiry.

We have already considered one type of path-dependent option in detail: the American option. This is clearly path-dependent since there is usually a finite probability of the option being exercised before expiry and thus ceasing to exist. This occurs if the asset price ever enters the range where it is optimal to exercise. In general, any option contract can specify either European or American exercise rights, and in this sense we can think of American early exercise rights as a feature of an option.

An American early exercise feature turns any option into a potentially path-dependent option.

Broadly speaking, an **exotic option** is an option that is not a vanilla put or call. Conceptually, the simplest exotic options are the **binary** or **digital** options. These are options that have payoffs that are more general than and different from from those of vanilla options. They may have either European or American style exercise features and have already been considered in Chapters 5 and 7. Originally the term 'binary' described an option that was, effectively, a straight bet on whether the underlying asset value would be above (a cash-or-nothing call) or below (a cash-or-nothing put) the exercise price; the payoff was independent of how far above or below the exercise price the asset value was at the time of exercise. Now, however, the term is used to describe any option with a payoff more general than the payoff for a put or a call.

Another relatively simple class of exotic options consists of **barrier options**. These are options where either the right to exercise is forfeited if the underlying asset value crosses a certain value (an **out barrier**), or the option comes into existence only if the asset value crosses a certain value (an **in barrier**). An example is a **down-and-out option**, where the right to exercise is lost if the asset price ever falls to or below some given down-and-out barrier. The option cannot be exercised thereafter and is worthless. (One reason for their existence is that they can be cheaper than their vanilla counterpart; another is that they may be a part of a more complicated structured contract.)

It is useful to think of barriers, whether in or out, simply as features of the option contract. In principle, a barrier can be applied to any option, whether a vanilla option, a binary or one of the class of path-dependent exotic options that we shall shortly describe.

A barrier feature makes an option path-dependent. If the asset price ever crosses an out barrier, the option becomes worthless and effectively ceases to exist. In the case of an in barrier, the option is worthless unless the asset price crosses the in barrier and the option to exercise does not exist until the in barrier is crossed.

An option with an American early exercise feature is path-dependent but not necessarily exotic. Similarly there are exotic options, such as binaries, that are exotic but not path-dependent. As suggested above, the term 'exotic' is loosely used to denote something out of the ordinary and usually not (currently) quoted on an exchange. Such options are usually traded over-the-counter, meaning that option brokers bring together both sides of a contract and construct a product which does not exist as an exchange-traded option; alternatively, one party may market and sell the option to clients. For example, while a vanilla American call is path-dependent, it is not considered exotic, whereas a European binary option is not path-dependent but it is considered exotic. There are, of course, exotic options which are also path-dependent.

The following is a list of some common exotic or path-dependent options which can all be put into the same framework and valued quite easily (albeit numerically, if necessary):

- binaries;
- compounds;
- choosers;
- barriers;
- Asians;
- lookbacks.

The first three of these are either not path-dependent at all or trivially so and we describe only the methodology behind their valuation. Barrier options, however, are discussed in detail in Chapter 12. The last two, Asians and lookbacks, are both crucially path-dependent (and from a mathematical point of view, particularly interesting); they each have a chapter to themselves, Chapters 14 and 15 respectively.

This is by no means an exhaustive list of exotic options. The number of options is continuing to grow rapidly and now includes range forwards, ladders, exchange, two-colour rainbow and cliquet, among others. Some of these (for example, the rainbow and exchange) depend on the values of many underlying assets, rather than a single underlying asset. In this and the following four chapters we aim to address questions that are of fundamental importance in the modelling and analysis of exotic options rather than to catalogue solutions for the more esoteric options. Any such catalogue would in any case quickly become out of date.

11.2 Compound Options: Options on Options

A **compound option** may be described simply as an option on an option. We consider only the case where the underlying option is a vanilla put or call and the compound option can be described as a vanilla put or call on the underlying option. For simplicity we assume that both the compound option and the underlying option are both European. The extension to more complicated path-dependent or exotic features for a compound option written on exotic or path-dependent underlying options is relatively straightforward.

Since there are two types of vanilla options, calls and puts, we can construct four different classes of basic compound option: calls on calls, calls on puts and so forth. We now consider a particular case, a call-on-a-call.

Let T_1 be the time at which we can, if we wish, exercise the compound call option and purchase the underlying vanilla call option for an amount E_1. This underlying call option may be exercised at time T_2 for an amount E_2 in return for an asset with price S. This a call-on-a-call, whilst superficially appearing complicated, may be valued very elegantly within the Black–Scholes framework we have met in earlier chapters.

The time interval that we need to consider in order to value the compound option is divided into two parts. Working back from expiry at $t = T_2$, we first find the value of the vanilla call option that we receive if we do, in fact, exercise the compound option at time $t = T_1$. This underlying call option has exercise price E_2 and expiry date T_2. There is an explicit formula for its value; even if there were not, we could find the solution by numerical means (so we could consider an underlying American option, for example). Thus, at time T_1, we can calculate the value of the underlying call option; let us call this $C(S, T_1)$. If, at time T_1, the asset price is such that $C(S, T_1) > E_1$ then we would clearly exercise our compound option and obtain the underlying call. If, however, at time T_1 the asset price is such that $C(S, T_1) < E_1$ then we would not exercise the compound option. Thus the payoff for the compound option at time T_1 is

$$\max\Big(C(S, T_1) - E_1, 0\Big). \tag{11.1}$$

Because the compound option's value is governed only by the random walk of the underlying asset price S, it too must satisfy the Black–Scholes equation. The only difference from a vanilla option is in the final condition: we use (11.1) as the final data in solving for the compound option for $t < T_1$.

The method of solution is similar for calls on puts, puts on calls and puts on puts and, in principle, exotics on exotics. American features do not change this solution strategy in any way other than in introducing constraints.

11.3 Chooser Options

Chooser options or **as-you-like-it options** are only slightly more complicated than compound options. Although they are, strictly speaking, path-dependent they can still be valued by solving the Black–Scholes equation.

A **regular chooser option** gives its owner the right to purchase, for an amount E_1 at time T_1, *either* a call *or* a put with exercise price E_2 at time T_2.Thus a regular chooser algorithm is a 'call on a call or put.' More general structures can readily be imagined, and presents no serious difficulties.

These options are valued in a manner similar to compound options. Again we assume that the options are all European. First we solve the underlying option problems; there are now two of these, one for the underlying call and one for the underlying put. We denote these solutions by $C(S,t)$ and $P(S,t)$ respectively, and use them as the final data for the chooser option problem. Clearly, one will exercise the first option if either $C(S,t) > E_1$ or $P(S,t) > E_1$, and one will elect to purchase the more valuable of the two. The chooser option again satisfies the Black–Scholes equation with final data at time T_1 given by

$$\max\Big(C(S,T_1) - E_1, P(S,T_1) - E_1, 0\Big). \qquad (11.2)$$

The contract can be made much more general than this by having the underlying call and put with different exercise prices and expiry dates, or by allowing the right to sell the vanilla put or call. Such a contract is called a **complex chooser option**.

11.4 Barrier Options

We devote the whole of Chapter 10 to **barrier options** and so we describe them only in outline here. These options are only weakly path-dependent in that they can be valued using only the current values of S and t, without the need for any variable to represent the path-dependent quantity; they therefore satisfy the Black–Scholes equation.[1] Barrier options differ from vanilla options in that part of the option contract is triggered if the asset price hits some barrier, $S = X$, say, at any time

[1] We are only considering vanilla options with barrier features here. There is no reason why a barrier feature cannot be applied to any option, whether vanilla or exotic. The underlying principles are exactly the same, since the barrier features affect only the boundary conditions in a partial differential equation formulation.

prior to expiry. As well as being either calls or puts, barrier options are categorised as follows:

- **up-and-in**: the option expires worthless *unless* the barrier $S = X$ is reached from *below* before expiry;
- **down-and-in**: the option expires worthless *unless* the barrier $S = X$ is reached from *above* before expiry;
- **up-and-out**: the option expires worthless *if* the barrier $S = X$ is reached from *below* before expiry;
- **down-and-out**: the option expires worthless *if* the barrier $S = X$ is reached from *above* before expiry.

Some barrier options specify a **rebate**, usually a fixed amount paid to the holder if the barrier is reached in the case of out-barriers or not reached in the case of in-barriers.

11.5 Asian Options

Asian options are the first fully path-dependent exotic options that we consider. They have payoffs which depend on the history of the random walk of the asset price via some sort of average. One such option is the **average strike call**, whose payoff is the difference between the asset price at expiry and its average over some period prior to expiry if this difference is positive, and zero otherwise.

Several factors affect the definition of average. Among these are:

- The period of averaging. Over what period prior to expiry is the average taken?
- Arithmetic or geometric averaging. The average can be defined as the mean of the asset price (the arithmetic average) or the exponential of the mean of the logarithm of the asset price (the geometric average).
- Weighted or unweighted averaging. Is the average simply the mean of asset prices over the averaging period or are some prices given a greater weighting in the average? We might, for example, choose to give a greater weighting to recent prices.
- Discrete or continuous sampling of the asset price. It is easier to take the mean of a small number of asset prices rather than the average over all realised asset prices. The average might, for example, be the mean of the closing asset prices at the end of every week before expiry instead of the average of the asset price measured every tick.

Different choices lead to different values for options. This list covers most of those used in practice.

When calculating the continuous mean of an asset price as it proceeds along its random walk we inevitably need to calculate time integrals of path-dependent quantities. We see in Chapter 13 that such integrals are of great importance when it comes to deriving a partial differential equation for the value of an option. The partial differential equation we find for the value of a continuously sampled Asian option contains one more term than the Black–Scholes equation. In contrast, when an average is measured from only a discrete sample of prices we find that the value of an option satisfies the Black–Scholes equation but now with jump conditions across the sampling dates. There is a very close analogy to be drawn with the earlier discussion of discretely paid dividends. This matter is again addressed in Chapter 13. Whether averages are measured continuously or discretely, the option valuation problem is now three-dimensional; we must keep track of the asset price, time, and now also the path-dependent quantity, which for Asian options is a running average.

In Chapter 13 we find a unifying framework for valuing many types of path-dependent option. Then we devote the whole of Chapter 14 to the analysis and valuation of Asian options in particular.

11.6 Lookback Options

A **lookback option** has a payoff that depends not only on the asset price at expiry but also on the maximum or the minimum of the asset price over some period prior to expiry. Usually the payoffs are structurally very similar to those of vanilla options. For example, a put option may have payoff

$$\max(J - S, 0)$$

where J is a suitably defined maximum. As with Asian options we can distinguish between discrete and continuous sampling of the asset price to obtain the maximum. Lookback options are discussed fully in Chapter 15 .

Further Reading

- Ingersoll (1987) discusses the partial differential equation approach to valuing some exotic options.

- For a treatment of compound options see Geske (1979). For options contingent on *two* assets see Stulz (1982).
- The general framework for path-dependent options is discussed in Dewynne & Wilmott (1993a).
- Some options depend on the values of many underlying assets, rather than a single underlying asset; see Barrett, Moore & Wilmott (1992).
- Babbs (1992) considers a binomial method for valuing lookback style options.
- Wilmott *et al.* (1993) describe the use of recursive binary trees to value path-dependent options.
- Information about practical issues, such as hedging of exotics, is given in the compilation of articles *From Black–Scholes to Black Holes*, published by *Risk* magazine (1993).

Exercises

1. What is the put-call parity result for compound options?

2. One might approach the compound option by considering the underlying option as the asset on which the compound option depends. Why is this not a good idea from the point of view of valuation?

3. It is possible to derive explicit formulæ for European compound and chooser options. With $C_{BS}(S, T_1)$ as the value at time T_1 of a European call option with strike price E_2 that expires on T_2, write the payoff function for a call on a call explicitly in terms of S. Sketch this payoff as a function of the underlying asset. Now use formula (5.16) to give an explicit expression for the value of the compound option. Simplify this expression using the function for the cumulative distribution of a bivariate normal distribution.

4. Repeat the analysis of Exercise 3 for a chooser option.

5. What happens to a chooser option if $T_1 = T_2$?

6. The chooser option with $E_1 = 0$ has a particularly simple value: what is it?

7. In this book we see many options whose value depends on an asset path. Examples are the Asian option, whose payoff depends on an average, and the lookback, whose payoff depends on the maximum or minimum of the asset. Why might these payoffs be important to a client? Why would they want options with such characteristics? What other properties of an asset price random walk may be important for

certain clients? Invent new types of option, having path-dependent payoffs, that might be commercially viable.

8. Discuss how one could define and value an American compound option, that is, one where the underlying option could be bought at any time between the initiation of the compound option, $t = 0$, and its expiry $t = T_1$. What happens if the underlying option is American and is exercised?

12 Barrier Options

12.1 Introduction

For our first in-depth discussion of a path-dependent option we consider a vanilla barrier option. As mentioned in the previous chapter, the four basic forms of these options are 'down-and-out', 'down-and-in', 'up-and-out' and 'up-and-in'. That is, the right to exercise either appears ('in') or disappears ('out') on some boundary in (S, t) space, above ('up') or below ('down') the asset price at the time the option is created. An example is a European option whose value becomes zero if the asset price ever goes as low as $S = X$. If the payoff is otherwise the same as that for a call option then we call this product a European 'down-and-out' call. An 'up-and-out' has similar characteristics except that it becomes worthless if the asset price ever exceeds a prescribed amount. These options can be further complicated by making the position of the knockout boundary a function of time and by having a rebate if the barrier is crossed. In the latter case the holder of the option receives a specified amount Z if the barrier is crossed in the case of a 'down' option or never crossed in the case of an 'in' option; this can make the option more attractive to potential purchasers.

We discuss only European options in any detail and we find a number of explicit formulæ for the values of various barrier options. The problem can be readily generalised to incorporate early exercise, although we must then find solutions numerically. In principle, barrier features may be applied to any options.

12.2 Knock-outs

We first consider the case of a European style down-and-out call option with payoff at expiry of $\max(S-E,0)$, provided that S never falls below X during the life of the option. If S ever reaches X then the option becomes worthless. This option has an explicit formula for its fair value. We consider in detail only the case where $E > X$.

For as long as S is greater than X, the value of the option $V(S,t)$ satisfies the Black–Scholes equation (3.9). As before, the final condition for this equation is

$$V(S,T) = \max(S - E, 0).$$

As S becomes large the likelihood of the barrier being activated becomes negligible and so assuming no dividends are paid

$$V(S,t) \sim S \quad \text{as} \quad S \to \infty.$$

So far, the problem is identical to that for a vanilla call. However, the valuation problem differs in that the second boundary condition is applied at $S = X$ rather than at $S = 0$. If S ever reaches X then the option is worthless; thus on the line $S = X$ the value of the option is zero:

$$V(X,t) = 0.$$

This completes the formulation of the problem; we now find the explicit solution.

We use the change of variables first introduced in Section 5.3. That is, we let

$$S = Ee^x, \quad t = T - \tau/\tfrac{1}{2}\sigma^2, \quad V = Ee^{\alpha x + \beta\tau}u(x,\tau)$$

with $\alpha = -\frac{1}{2}(k-1)$, $\beta = -\frac{1}{4}(k+1)^2$ and $k = r/\frac{1}{2}\sigma^2$. In these new variables the barrier transforms to

$$x_0 = \log(X/E),$$

and the barrier option problem becomes

$$\frac{\partial u}{\partial \tau} = \frac{\partial^2 u}{\partial x^2} \tag{12.1}$$

with

$$u(x,0) = \max\left(e^{\frac{1}{2}(k+1)x} - e^{\frac{1}{2}(k-1)x}, 0\right) = u_0(x), \quad x \geq x_0, \tag{12.2}$$

$$u(x,t) \sim e^{(1-\alpha)x - \beta\tau} \quad \text{as} \quad x \to \infty, \tag{12.3}$$

and

$$u(x_0, t) = 0. \tag{12.4}$$

The last boundary condition is new and we deal with it by the **method of images**.

We have several times related the problem of valuing simple call and put options to the flow of heat in an infinite bar. Boundary condition (12.4) is, however, imposed at a finite value of x: the analogy is now with heat flow in a *semi*-infinite bar held at zero temperature at the point $x = \log(X/E)$.

The flow of heat in a bar is unaffected by the coordinate system used, so equation (12.4) is invariant under translation, from x to $x + x_0$, or reflection, from x to $-x$. Thus, if $u(x, \tau)$ is a solution of (12.1), so are $u(x + x_0, \tau)$ and $u(-x + x_0, \tau)$ for any constant x_0. In the method of images we solve a semi-infinite problem by first solving an infinite problem made up of *two* semi-infinite problems with equal and opposite initial temperature distributions: one half is hot, the other cold. The net effect is cancellation at the join: the temperature there is guaranteed to be zero.

We can apply this method to the barrier option problem. We reflect the initial data about the point $x_0 = \log(X/E)$ (the 'join' of the two bars), at the same time changing its sign, thereby automatically satisfying (12.4). Thus, instead of solving (12.1)–(12.4) on the interval $x_0 < x < \infty$, we solve (12.1) for all x but subject to

$$u(x, 0) = u_0(x) - u_0(2x_0 - x),$$

that is,

$$u(x,0) = \begin{cases} \max\left(e^{\frac{1}{2}(k+1)x} - e^{\frac{1}{2}(k-1)x}, 0\right) & \text{for } x > x_0 \\ -\max\left(e^{(k+1)(x_0 - \frac{1}{2}x)} - e^{(k-1)(x_0 - \frac{1}{2}x)}, 0\right) & \text{for } x < x_0. \end{cases}$$

In this way we guarantee that $u(x_0, 0) = 0$.

Suppose that

$$C(S, t) = Ee^{\alpha x + \beta \tau} u_1(x, \tau) \tag{12.5}$$

is the value of a vanilla option with the same exercise price and expiry date but with no barrier. Of course, this value is given by the Black–Scholes formula and we know that $u_1(x, \tau)$ is a solution of the heat equation. Thus, we have

$$u_1(x, \tau) = e^{-\alpha x - \beta \tau} C(S, t)/E.$$

Next we write the solution to the barrier option value as

$$V(S,t) = Ee^{\alpha x+\beta\tau}\left(u_1(x,\tau) + u_2(x,\tau)\right)$$

where $u_2(x,\tau)$ is the solution of the problem with antisymmetric initial data. The solution of this problem can be found in terms of u_1 by using the invariance of the equation (12.1) under translation and changes of sign. We must have

$$\begin{aligned} u_2(x,\tau) &= -u_1(2x_0 - x,\tau) \\ &= -e^{-\alpha(2\log(X/E)-\log(S/E))-\beta\tau}C(X^2/S,t)/E, \end{aligned}$$

since replacing x by $2x_0-x$ is equivalent to replacing S by X^2/S. Finally, bringing all these together and writing the solution purely in terms of S and t, we have

$$V(S,t) = C(S,t) - \left(\frac{S}{X}\right)^{-(k-1)} C(X^2/S,t).$$

It is obvious that $V(X,t) = 0$; it can also be verified that the equation and final condition are also satisfied. (The final condition is satisfied only for $S > X$, of course; for $S < X$ the option is worthless.)

This demonstrates the use of the method of images to find an explicit formula for a down-and-out call option. Other 'out' options can be valued similarly and are left as exercises. However, 'in' options must be treated slightly differently.

12.3 Knock-ins

An 'in' option expires worthless *unless* the asset price reaches the barrier before expiry. If the asset value crosses the line $S = X$ at some time prior to expiry then the option becomes a vanilla option with the appropriate payoff. It is common for in-type barrier options to give a rebate, usually a fixed amount, if the barrier is not hit. This compensates the holder for the loss of the option.

Let us now consider a down-and-in European call option. The option value $V(S,t)$ still satisfies the basic Black–Scholes equation (3.9), and all we have to do to pose the problem fully is to determine the correct final and boundary conditions. We use the notation $C(S,t)$ to denote the European vanilla call with the same expiry date and exercise price as the barrier call.

Once the barrier has been crossed, the option is a vanilla call whose value is given by the Black–Scholes formula. Let us now consider the situation in which the barrier has yet to be crossed.

The option is worthless as $S \to \infty$. This is because the larger that S is, the less likely it is to fall through the barrier before expiry and activate the option. Thus one boundary condition is

$$V(S,t) \to 0 \quad \text{as} \quad S \to \infty.$$

If S has been greater than X right up to expiry then the option expires worthless. The final condition, for $S > X$, is therefore[1]

$$V(S,T) = 0.$$

Finally, should the asset price reach $S = X$ at some time before expiry the option immediately turns into a vanilla call and must thus have the same value as this call. The second boundary condition is therefore

$$V(X,t) = C(X,t).$$

If $S < X$ at any point then the barrier has been crossed, the option is activated and the value of the option is exactly the same as a vanilla call. Thus, we have only to solve for the value in $S > X$. This completes the formulation of the European down-and-in barrier call.

In order to solve the down-and-in explicitly we first write

$$V(S,t) = C(S,t) - \bar{V}(S,t).$$

Since the Black–Scholes equation and boundary conditions are linear we know that $\bar{V}$ must satisfy the Black–Scholes equation with final condition

$$\bar{V}(S,T) = C(S,T) - V(S,T) = C(S,T) = \max(S - E, 0);$$

and boundary conditions

$$\begin{aligned} \bar{V}(S,t) &= C(S,t) - V(S,t) \sim S - 0 = S, \quad \text{as } S \to \infty \\ \bar{V}(X,t) &= C(X,t) - V(X,t) = C(X,t) - C(X,t) = 0. \end{aligned}$$

This is the problem for the down-and-*out* barrier option. In other words, in this case, a European 'in' plus a European 'out' equals a vanilla. This is obvious from a financial point of view, as the value of a portfolio consisting of one in-option and one out-option (with the same barrier, exercise price and expiry dates) is obviously equal to the value of a vanilla call (with the same exercise price and expiry dates). This is

[1] If the option contract specifies a rebate Z if the barrier is never crossed this becomes $V(S,T) = Z$.

because only one of the two barrier options can be active at expiry and whichever it is, its value is the value of a vanilla call.

The American versions of all the barrier options exist but do not in general have explicit formulæ. Nevertheless, their numerical solution is no harder than for vanilla options.

Further Reading

- Rubinstein (1992) contains a catalogue and explicit formulæ for a large number of barrier options.

Exercises

1. How is an 'out' boundary condition changed if an out-option pays a rebate of Z if the barrier is triggered?

2. How is an 'in' final condition changed if a rebate of Z is paid if the barrier is never triggered?

3. Find explicit formulæ for all varieties of European barrier options (in/out, up/down, call/put) including a rebate.

4. Describe the evolution of the delta hedge of written down-and-in and down-and-out barrier call options, considering separately the cases in which the barrier is and is not triggered. (Compare with Exercise 5 in Chapter 3.)

5. The **double knockout** call or put option expires worthless if the asset price *either* rises to an upper barrier value X_2 *or* falls to a lower barrier X_1; otherwise its payoff is that of a vanilla call or put. What is its value?

6. By seeking solutions of the Black–Scholes equation which are independent of time, show that there are 'perpetual' barrier options, i.e. ones whose values are independent of t. These options have no expiry date ($T = \infty$). Find their explicit formulæ and include a continuously paid constant dividend yield on the underlying.

7. How would you value a barrier option that pays \$1 if an asset price first rises to some given level X_0 *and then* falls to another level X_1 before a time T, and otherwise pays nothing?

8. If $V(S,t)$ satisfies the Black–Scholes equation, show that for any constants α and a, $U(S,t) = S^\alpha V(a/S,t)$ satisfies

$$\frac{\partial U}{\partial t} + \tfrac{1}{2}\sigma^2 S^2 \frac{\partial^2 U}{\partial S^2} - (r + (\alpha - 1)\sigma^2) S \frac{\partial U}{\partial S} - (1-\alpha)\left(r + \tfrac{1}{2}\sigma^2\right) U = 0.$$

What is special about the case $\alpha = 1 - r/\frac{1}{2}\sigma^2$? Use this result to find the formulæ for the values of the down-and-out option and the up-and-out option.

9. The 'bouncing ball option' pays out if the asset price crosses a barrier three times, say, during the life of the option. This description, as it stands, is meaningless. Why? Define your own bouncing ball option, with a properly specified, and meaningful, payoff. How would you value your option?

10. What changes would be necessary to the boundary and initial conditions for the explicit finite-difference method to value a down-and-out call option (with zero rebate)? What modifications would be needed to the pseudo-code in Figure 8.5?

11. How could you modify the binomial methods described in Chapter 10 to deal with barrier options?

13 A Unifying Framework for Path-dependent Options

13.1 Introduction

When we come to analyse options that depend on a path-dependent quantity, such as an average of the asset price, the straightforward Black–Scholes approach, which has hitherto stood us in good stead, is inadequate. The reason is simply that, although there are many realisations of the asset price's random walk leading to the current value, in general any two of these give a different value for the path-dependent quantity. We are therefore led to the idea of introducing a third independent variable, in addition to S and t, whose rôle is to measure the relevant path-dependent quantity. This idea leads to a general framework within which we can handle such seemingly disparate options as Asian options (depending on an average of asset values) and lookback options (depending on the realised maximum or minimum of the asset price). In this chapter we introduce the framework; it is put into practice in Chapters 14 (Asian options) and 15 (lookbacks).

To introduce the idea let us consider the form of the payoff for an average strike option. For the first part of this chapter we consider continuously sampled quantities. That means that our average will depend on a time integral. In the case of the **arithmetic-average strike call option** we have a payoff of the form

$$\max\left(S - \frac{1}{T}\int_0^T S(\tau)\,d\tau\right).$$

What does the integral term mean here, and why is an option with this kind of payoff called an average strike option?

First, observe that the continuously measured arithmetic average of an asset price over the period 0 to T is simply the integral of the asset price as a function of time, divided by the duration of the period, T (we see other forms of average later). Such a contract is called an average strike option because the rôle played by the strike or exercise price in a vanilla option is here taken by the average.

13.2 Time Integrals of the Random Walk

Motivated by the example above, let us consider a fairly general class of European options with payoff depending on S and on

$$\int_0^T f\left(S(\tau), \tau\right) d\tau, \tag{13.1}$$

where f is a given function of the variables S and t. The integral in (13.1) is over the path of S from the initiation of averaging at $t = 0$ to expiry at $t = T$. For the average strike call option, where the payoff at expiry is

$$\max\left(S - \frac{1}{T}\int_0^T S(\tau)\, d\tau, 0\right),$$

we have $f(S,t) = S$.

We introduce the new variable

$$I = \int_0^t f\left(S(\tau), \tau\right) d\tau. \tag{13.2}$$

Since the history of the asset price is independent of the current price, we may treat I, S and t as independent variables; different realisations of the random walk lead to different values of I. Observe that this definition (13.2) is simply (13.1) with the expiry date T replaced by t. Because the payoff depends on both I and S, we anticipate that the value of an exotic path-dependent option can be written as $V(S,I,t)$. That is, the option value is a function of *three* independent variables: time t, the current asset price S and the history integral of the asset price I.

We intend applying Itô's lemma to V, and to do this we need to know the stochastic differential equation for I. This is found quite easily by considering the change in I as t and S change by small amounts. Clearly,

$$I(t + dt) = I + dI = \int_0^{t+dt} f\left(S(\tau), \tau\right) d\tau.$$

To $O(dt)$ this can be written as

$$I + dI = \int_0^t f\left(S(\tau), \tau\right) d\tau + f\left(S(t), t\right) \, dt,$$

so that

$$dI = f(S,t)\, dt. \tag{13.3}$$

This is the stochastic differential equation for I; it so happens that there is no random component. We are now in a position to value any option that depends on S, t and I.

First we apply Itô's lemma to the function $V(S, I, t)$ to show that

$$dV = \sigma S \frac{\partial V}{\partial S} dX + \left(\tfrac{1}{2}\sigma^2 S^2 \frac{\partial^2 V}{\partial S^2} + \mu S \frac{\partial V}{\partial S} + \frac{\partial V}{\partial t} + f(S,t) \frac{\partial V}{\partial I} \right) dt. \tag{13.4}$$

This is derived in exactly the same way as equation (3.3). Note that the new term, which is proportional to the rate of change of V with respect to I, does not introduce any new *stochastic* terms into the random walk followed by V. (Since dI introduces no new source of risk, we anticipate that the option can be hedged using the underlying only.)

Recalling that the option is European, we now set up the usual risk-free portfolio, consisting of one option and a short position with a number Δ of the underlying. The delta is still equal to $\partial V / \partial S$, and we find that arbitrage considerations lead to

$$\frac{\partial V}{\partial t} + f(S,t) \frac{\partial V}{\partial I} + \tfrac{1}{2}\sigma^2 S^2 \frac{\partial^2 V}{\partial S^2} + rS \frac{\partial V}{\partial S} - rV = 0. \tag{13.5}$$

Observe that this equation is identical to the basic Black–Scholes equation except that there is one extra term, the derivative of V with respect to I.

As for all derivatives, we solve the equation with a final condition. At expiry we know the exact form of the payoff and hence the option value as a function of S and I. We have

$$V(S, I, T) = \Lambda(S, I, T)$$

where the function Λ is the known payoff function. In the case of the average strike call we would take $I = \int_0^t S(\tau)\, d\tau$ and then

$$\Lambda(S, I, T) = \max(S - I/T, 0).$$

Technical Point: Early Exercise.
We can easily extend the analysis to American options. Suppose that we wish to value an American version of the average strike option. In any such contract the payoff on early exercise must be specified in advance. Let us suppose that the early exercise payoff for this average strike call is

$$\max(S - I/t, 0).$$

This is a natural choice for this particular option, since it depends on the asset price average to date, the running average. For some options, especially those depending on discretely measured path-dependent quantities, the payoff is not so obvious. Nevertheless, let us suppose that in our general framework the payoff takes the form

$$\Lambda(S, I, t),$$

where Λ is a function known in advance and specified in the option contract.

As in Chapter 3 for American vanilla options, the American path-dependent option valuation problem is a simple modification of the European case. To this end, we introduce the partial differential operator

$$\mathcal{L}_{EX} = \frac{\partial}{\partial t} + f(S,t)\frac{\partial}{\partial I} + \tfrac{1}{2}\sigma^2 S^2 \frac{\partial^2}{\partial S^2} + rS\frac{\partial}{\partial S} - r.$$

This operator is a generalisation of the Black–Scholes operator. It measures the difference between the rates of return on a risk-free delta-hedged portfolio and a bank deposit of equivalent value. As for the vanilla option case, the rate of return from a delta-hedged portfolio cannot exceed the rate of return from a bank deposit, but it need not equal it as there may be times when it is optimal to exercise the option early. Thus

$$\mathcal{L}_{EX}(V) \leq 0.$$

Arbitrage considerations show that we must always have

$$V(S, I, t) \geq \Lambda(S, I, t).$$

If the rate of return from the portfolio equals the rate of return from a bank deposit, $\mathcal{L}_{EX}(V) = 0$, then it is not optimal to exercise the option. This can be the case only if it is more valuable held or sold than exercised, $V > \Lambda$. If the rate of return from the portfolio is less than the return from a bank deposit, $\mathcal{L}_{EX}(V) < 0$, then it is optimal to exercise the option. This can be the case only if $V = \Lambda$ (if $V > \Lambda$ then it would be more profitable to sell the option, and we can never have $V < \Lambda$). Thus either $\mathcal{L}_{EX}(V) = 0$ and $V - \Lambda > 0$ or $\mathcal{L}_{EX}(V) < 0$ and $V - \Lambda = 0$. Either way, we always have

$$\mathcal{L}_{EX}(V) \cdot (V - \Lambda) = 0.$$

Thus the problem for the American version of our class of path-dependent exotic options can be written in linear complementarity form as

$$\mathcal{L}_{EX}(V) \cdot (V - \Lambda) = 0, \quad \mathcal{L}_{EX}(V) \leq 0, \quad (V - \Lambda) \geq 0, \tag{13.6}$$

with V and $\partial V/\partial S$ continuous (assuming Λ is continuous) and with final condition

$$V(S, I, T) = \Lambda(S, I, T).$$

The condition that the delta, i.e. the derivative of V with respect to S, must always be continuous follows from the same arbitrage argument as earlier. This assumes, as stated above, that the payoff function $\Lambda(S, I, t)$ is continuous in S.

13.3 Discrete Sampling

We hinted earlier that when path-dependent quantities are measured using a finite sample of asset prices we no longer have a partial differential equation with a new derivative with respect to a new variable. Instead, we simply solve the basic Black–Scholes equation with jump conditions across the sampling dates, much as when the asset pays discrete dividends. We must still solve in three dimensions as above but we can treat the extra path-dependent quantity merely as a parameter in the problem.

The best way to introduce the basic idea is to consider a simple case (it is covered in detail in Chapter 14). Consider an Asian option with a payoff that depends on a discretely measured arithmetic average of the realised asset prices, i.e. it depends on

$$\frac{1}{N} \sum_{i=1}^{N} S(t_i).$$

Here the t_i are the N **sampling dates**. It would then be sensible to assume that the value of the option depends on I where

$$I = \sum_{i=1}^{j(t)} S(t_i), \tag{13.7}$$

and where $j(t)$ is the largest integer such that $t_{j(t)} \leq t$. Thus $V = V(S, I, t)$. There is an obvious analogy between the representation of a continuous sampling as an *integral* and discrete sampling as a *sum*.

In Chapter 6 we demonstrated how to allow for discretely paid dividends in the (continuous) Black–Scholes model. The analysis of that chapter showed how a simple financial argument led to jump conditions at the dividend date. The same analysis is possible when path-dependent quantities are sampled discretely and again results in a jump condition.

In the case of the average strike call with the definition (13.7) for the running sum, we find that, across a sampling date t_i, the sum is updated from I before the sampling date to a new value just after. This new value is just the sum of the old value and the value of S at the date t_i. In other words, I is updated across a sampling date by the simple rule

$$\text{new value of } I = \text{old value of } I + S.$$

We now ask the question:

- Does the value of the option jump across the sampling date?

The answer to this question is, as for the case of discrete dividend payments, both yes and no, depending on the way that the option value is viewed in relation to the underlying. It is certainly true that $V(S, I, t)$ need not be continuous *for fixed S and I* as t varies. In that sense, there is indeed a jump in V, and the answer to the question is 'yes'. However, in the course of any realisation of the asset price in which all of S, I and t vary, the option price does *not* change discontinuously, and the answer is 'no'. This latter statement is a simple consequence of the absence of arbitrage opportunities: if the value of an option jumped discontinuously across a known sampling date it would present an obvious arbitrage opportunity.

These two, apparently contradictory, statements can be reconciled once it is recognised that across a sampling date the discretely sampled average changes discontinuously because I is measured discretely. The discontinuity of I and the continuity of $V(S(t), I(t), t)$ for any *realisation* of the random walk forces $V(S, I, t)$ (viewed as a function of t with S and I fixed) to change discontinuously across sampling dates.

We introduce the notation I_i to denote the value of the running sum I for $t_i < t < t_{i+1}$, and S_i for the value of S at the sampling date t_i. Thus, I_i represents the (constant) value of I for the period immediately after a sample taken at t_i until the next sample is taken at t_{i+1}. We may therefore write the updating rule as

$$I_i = I_{i-1} + S_i, \tag{13.8}$$

so I is updated at time t_i by adding to it the value of S at that time. Since I_i is constant for the period t_i^+ (just after a sample is taken) to t_{i+1}^- (immediately before the next sample), it is effectively a parameter in the value of the option during this time, in the same way that the exercise price is a parameter in the value of a vanilla option, or the dividend rate is when dividends are paid discretely. During this period, the only random variable that is changing is S and the option price must therefore satisfy the basic Black–Scholes equation during this time. From (13.8) it is clear that I is discontinuous at t_i as we noted above. However, since the *realised* option price is continuous across t_i we have

$$V(S_i, I_i, t_i^+) = V(S_i, I_{i-1}, t_i^-), \tag{13.9}$$

where I_{i-1} is the value of I immediately before sampling and I_i is the value immediately after sampling. Of course, for each realisation S is continuous and takes the same value immediately before and immediately after sampling. Using (13.8), (13.9) can be written as

$$V(S, I_{i-1} + S, t_i^+) = V(S, I_{i-1}, t_i^-). \tag{13.10}$$

Since I_{i-1} does not change from t_{i-1}^+ to t_i^- we can drop its suffix $i-1$ in (13.10) with no possibility of confusion and arrive at the jump condition

$$V(S, I, t_i^-) = V(S, I + S, t_i^+). \tag{13.11}$$

This is the jump condition for the discretely sampled arithmetic Asian option. Notice that in (13.9) we think of S and I arising from a realisation of the random walk (so that they vary in time) and in (13.11) we think of them as fixed. Notice also that it is implicit in (13.11) that we solve backwards in time, from t_i^+ to t_i^-.

This derivation can be applied to any option which depends on a discretely updated parameter. For example, if the option depends on an I determined by a general equation of the form

$$I_i = w_i(S_i, I_{i-1}),$$

(where the functions w_i are known in advance) the jump condition is simply

$$V(S, I, t_i^-) = V\left(S, w_i(S, I), t_i^+\right).$$

In particular, for the discretely sampled geometric average where the relevant running sum is

$$I = \sum_{i=1}^{j(t)} \log S(t_i)$$

we find that

$$V(S, I, t_i^-) = V\left(S, I + \log S, t_i^+\right).$$

Another simple and important example is the lookback option, of which more in Chapter 15, where the maximum is updated with the rule

$$I_i = \max(I_{i-1}, S).$$

This leads to the jump condition

$$V(S, I, t_i^-) = V\left(S, \max(I, S), t_i^+\right).$$

We can see that, although the particular definition of the discrete average affects the details of the jump conditions across sampling dates, it does not affect the general procedure for solution. This is because the path-dependent quantity, I, is updated discretely and is therefore constant between sampling dates. The partial differential equation for the option value between sampling dates is just the basic Black–Scholes equation with I treated as a parameter. Thus the strategy for valuing any path-dependent option with discrete sampling is as follows:

- Starting from the expiry date, where the option value is known (equal to the payoff), and working backwards in time, solve

  $$\frac{\partial V}{\partial t} + \tfrac{1}{2}\sigma^2 S^2 \frac{\partial^2 V}{\partial S^2} + rS\frac{\partial V}{\partial S} - rV = 0$$

  between sampling dates, using the value of the option immediately before the next sampling date as final data. This gives the value of the option until immediately after the present sampling date.
- Then apply the appropriate jump condition across the current sampling date to deduce the option value immediately before the present sampling date.
- Repeat this process as necessary to arrive at the current value of the option.

Exercises

1. Use Dirac delta functions in the definition of the path-dependent quantity to bring together the mathematical derivation of the partial differential equation for exotic options and the financial derivation of the jump conditions for discrete sampling.

2. In exercise 7 of Chapter 11 you were asked to invent new types of path-dependent option. Can your options be put into the framework presented in this chapter? If so, how?

14 Asian Options: Options on Averages

14.1 Introduction

A typical example of an Asian option is a contract giving the holder the right to buy an asset for its average price over some prescribed period. Such a product is of obvious appeal to a company which must buy a commodity at a fixed time each year, yet has to sell it regularly throughout the year. In this case the underlying asset is the commodity. The same type of option is also used in foreign exchange markets by companies that have continuous sales in one currency but must purchase raw materials in a different currency and at a fixed date. Here, the underlying is the exchange rate. These options allow investors to insure against losses from adverse movements in an underlying asset without the need for continuous rehedging.

In this chapter we derive differential equations for the value of several Asian options. The common feature is that the exercise price is always some form of average of the price of the underlying over some period prior to exercise. The exercise price may depend on geometric or arithmetic averages, which may be measured either continuously or discretely. As well as deriving the equations we examine several problems in more detail, in particular the continuously sampled arithmetic average strike option with either European or American exercise features, and the European geometric average strike with either continuous or discrete sampling. In general Asian options depend on three independent variables (see Chapter 13), but we find that these particular options permit similarity reductions where the value of the option is in each reduced to a function of *two* variables. This enables us to derive explicit solutions for options depending on geometric averages. (These are not the only similarity solutions derivable for Asian options. Many

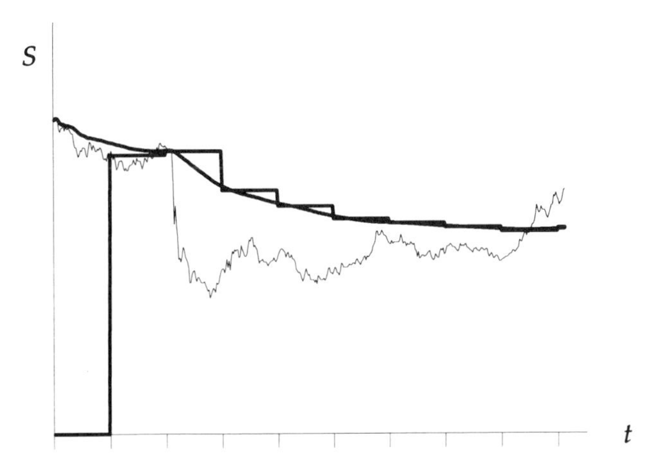

Figure 14.1 An asset price random walk, its continuously measured arithmetic running average and a discrete arithmetic running average.

other types of Asian option can be simplified and solved in this way, and these are left as exercises for the reader.)

More details about the analytical and numerical valuation of Asian options can be found in *Option Pricing*.

14.2 Continuously Sampled Averages

14.2.1 Arithmetic Averaging

Figure 14.1 shows a realisation of the random walk followed by an asset together with two versions of its running arithmetic average. One is the continuous arithmetic running average defined above, which is initiated at the start of the graph; the figure also shows the discrete version of this average, which is discussed later in this chapter.

The basic model for valuing Asian options is discussed in Chapter 13, and the general form of the partial differential equation governing the option value is equation (13.5). In particular, for an option depending on the continuously sampled arithmetic average

$$\frac{1}{t}\int_0^t S(\tau)\,d\tau,$$

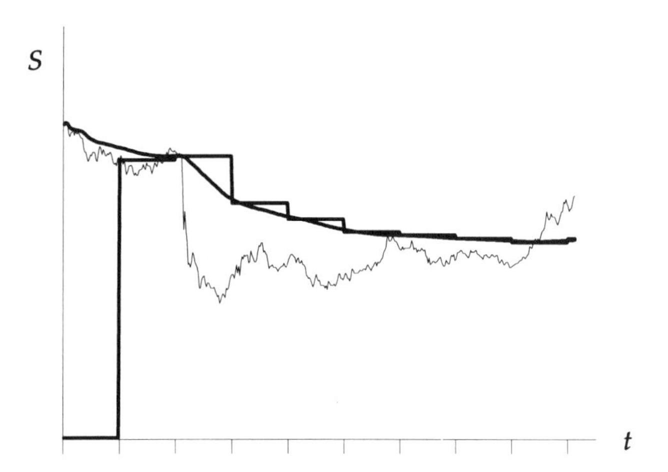

Figure 14.2 An asset price random walk, its continuous geometric running average and a discrete geometric running average.

we introduce the variable

$$I = \int_0^t S(\tau)\, d\tau. \tag{14.1}$$

Following the analysis of Section 13.2, the partial differential equation for the value of such an option is

$$\frac{\partial V}{\partial t} + S\frac{\partial V}{\partial I} + \tfrac{1}{2}\sigma^2 S^2 \frac{\partial^2 V}{\partial S^2} + rS\frac{\partial V}{\partial S} - rV = 0. \tag{14.2}$$

This follows as (14.1) is simply (13.1) with $f(S,\tau) = S$ and (14.2) is, accordingly, (13.5) with the same substitution.

14.2.2 Geometric Averaging

Figure 14.2 shows a realisation of an asset price random walk with the continuous and discrete versions of its geometric running average.

The continuously sampled geometric average[1] is defined to be

$$\exp\left(\frac{1}{t}\int_0^t \log S(\tau)\, d\tau\right);$$

[1] Strictly speaking, we should make S dimensionless in this formula; however, the additive constant that arises from the units of S always cancels out. We can think of S as dimensionless in the units of currency under consideration.

it is the limit as $n \to \infty$ of the discrete geometric average

$$\left(\prod_{i=1}^{n} S(t_i) \right)^{1/n}.$$

When this determines the payoff of the option, we define

$$I = \int_0^t \log S(\tau)\, d\tau$$

and, following again the analysis of Section 13.2, the partial differential equation for the value of the option is

$$\frac{\partial V}{\partial t} + \log S \frac{\partial V}{\partial I} + \tfrac{1}{2}\sigma^2 S^2 \frac{\partial^2 V}{\partial S^2} + rS \frac{\partial V}{\partial S} - rV = 0. \tag{14.3}$$

14.3 Similarity Reductions

The value of an Asian option depends on three variables S, I and t. This is true whether the quantity I is measured arithmetically or geometrically, continuously or discretely. Typically the value of these options must be calculated numerically. In cases where the option valuation problem is genuinely in three dimensions any computer program will be much slower than that for a vanilla option, due to the extra dimension. This cannot be avoided. However, some options have a particular mathematical structure that permits a reduction in the dimensionality of the problem by use of a similarity variable; we discuss two such options in detail in the following sections. We saw in Chapter 5 how a problem in two dimensions could be reduced to a problem in only one dimension because of the structure of the differential equation and its boundary and initial conditions. In the case of the arithmetically sampled Asian option we can reduce the problem from three to two dimensions when the following condition holds:

- the payoff has the form $S^\alpha F(I/S, t)$ for some constant α (usually equal to 0 or 1) and some function F.

If this condition holds we only need to find solutions of a two-dimensional problem. In fact, for continuous averaging, we find that $V = S^\alpha H(R, t)$, where $R = I/S$, and

$$\begin{aligned} &\frac{\partial H}{\partial t} + \tfrac{1}{2}\sigma^2 R^2 \frac{\partial^2 H}{\partial R^2} \\ &\quad + \left(1 + (\sigma^2(1-\alpha) - r)R\right) \frac{\partial H}{\partial R} - (1-\alpha)(\tfrac{1}{2}\sigma^2 \alpha + r)H = 0, \end{aligned} \tag{14.4}$$

with payoff

$$H(R, T) = F(R).$$

This problem in two variables is a lot less pleasant than the analogous Black–Scholes problem. Although it has a solution in terms of an infinite sum of confluent hypergeometric functions, it is usually more practical to solve the partial differential equation numerically than to evaluate the infinite sum. For an American style option, the similarity reduction $V = S^\alpha H(R, t)$, $R = I/S$ reduces the Black–Scholes inequality to (14.4), but with inequality rather than equality, i.e. with a $\leq$ replacing the $=$ sign.

14.4 The Continuously Sampled Average Strike Option

For our main example of an Asian option we examine in depth the average strike option, where the payoff at expiry is

$$\max\left(S - \frac{1}{T}\int_0^T S(\tau)\,d\tau,\, 0\right),$$

in which I is the running average evaluated at expiry, for a call. For a put the payoff is

$$\max\left(\frac{1}{T}\int_0^T S(\tau)\,d\tau - S,\, 0\right),$$

but we give details only for the call.

If we want to value the American version, we must first decide on the payoff for early exercise; the average up to expiry is not known before expiry. The natural choice for the call is

$$\max\left(S - \frac{1}{t}\int_0^t S(\tau)\,d\tau,\, 0\right); \tag{14.5}$$

this payoff depends on the running average from time 0 to t, and agrees with the payoff at expiry.

We can write the running payoff for the call option as

$$S\max\left(1 - \frac{1}{St}\int_0^t S(\tau)\,d\tau,\, 0\right),$$

which is consistent with the change of variables

$$R = \frac{1}{S}\int_0^t S(\tau)\,d\tau \tag{14.6}$$

that we introduced above. The payoffs for early exercise and at expiry may then be written respectively as

$$S \max(1 - R/t, 0), \quad S \max(1 - R/T, 0).$$

(Note that, written in terms of R, the payoff for the call option looks more like that of a *put*; if we had used S/I for the similarity variable it would have looked more like a call.) In view of the forms of the payoff functions above and the discussion in the previous section, we are led to postulate that the option value takes the form

$$V(S, R, t) = SH(R, t), \quad \text{with} \quad R = I/S;$$

in this case $\alpha = 1$. We find that

$$\frac{\partial H}{\partial t} + \tfrac{1}{2}\sigma^2 R^2 \frac{\partial^2 H}{\partial R^2} + (1 - rR)\frac{\partial H}{\partial R} \leq 0. \qquad (14.7)$$

If the option is European we have strict equality in (14.7). If it is American we may have inequality in (14.7) but the constraint

$$H(R, t) \geq \Lambda(R, t) = \max(1 - R/t, 0) \qquad (14.8)$$

must be satisfied. Moreover, if the option price ever meets the early exercise payoff it must do so smoothly. That is, the function $H(R, t)$ and its first R-derivative must be continuous everywhere.

Technical Point: Boundary Conditions for the European Option.

For the European option we must impose boundary conditions both at $R = 0$ and as $R \to \infty$. The boundary condition as $R \to \infty$ is simple. Since S is bounded for finite t, the only way that R can tend to infinity is for S to tend to zero. In this case the option will not be exercised, and so

$$H(\infty, t) = 0.$$

To determine the boundary condition at $R = 0$ we need to take a close look at the behaviour of R when it is small. From (14.6) we find that R satisfies the stochastic differential equation

$$dR = -\sigma R\, dX + \Big(1 + (\sigma^2 - \mu)R\Big)\, dt.$$

Now recall the discussion of the boundary condition at $S = 0$ for the vanilla option. This followed, since if ever S is zero then it remains zero for all time; the payoff is known with certainty. This, however, is not the case for the random walk in R. Put $R = 0$ in the above and we find that

$dR = dt > 0$, so the variable R immediately moves away from $R = 0$ into $R > 0$. Thus, even if $R = 0$ now, there is no reason why it should remain zero until expiry. Therefore we no longer know the value of the option when $R = 0$ with certainty. All we know is that the value of the option must be finite.

We can use the condition that the option value is finite at $R = 0$ to deduce a boundary condition there from the differential equation. First, note that for R small the term $R\partial H/\partial R$ is negligible compared with $\partial H/\partial R$. We can thus ignore this term (in fact, this is independent of whether H is finite or not). We may also ignore the $R^2\partial^2 H/\partial R^2$ term as $R \to 0$ for the following reason. Suppose that $R^2\partial^2 H/\partial R^2$ tends to a nonzero limit as $R \to 0$; we can assume without loss of generality that

$$\lim_{R\to 0} R^2 \frac{\partial^2 H}{\partial R^2} = O(1).$$

For small R we would then have

$$\frac{\partial^2 H}{\partial R^2} = O\left(\frac{1}{R^2}\right),$$

which may easily be integrated to show that $H = O(\log R)$ as $R \to 0$. This is, of course, inconsistent with H being finite. Thus we conclude that only the terms $\partial H/\partial t$ and $\partial H/\partial R$ can contribute near $R = 0$. In other words

$$\frac{\partial H}{\partial t} + \frac{\partial H}{\partial R} = 0 \quad \text{on} \quad R = 0. \tag{14.9}$$

This is the second boundary condition.[2]

The equation (14.7), with a final condition and boundary conditions at $R = 0$ and $R = \infty$, are sufficient to determine the value of a European option uniquely. As mentioned above, it is possible to write down an exact analytic expression for the problem as an infinite sum of confluent hypergeometric functions. We do not give this exact solution because the confluent hypergeometric function is not a widely known function, the solution is represented in terms of an infinite sum and because, from a practical point of view, it is quicker to obtain values by

[2] Other balances are possible near $R = 0$, for example

$$\frac{\partial H}{\partial R} + \tfrac{1}{2}\sigma^2 R^2 \frac{\partial^2 H}{\partial R^2} \sim 0.$$

This, and other balances, either are inconsistent with the equation (i.e. ignored terms are, in fact, not small), or lead to exponentially large option prices. The latter are financially unrealistic.

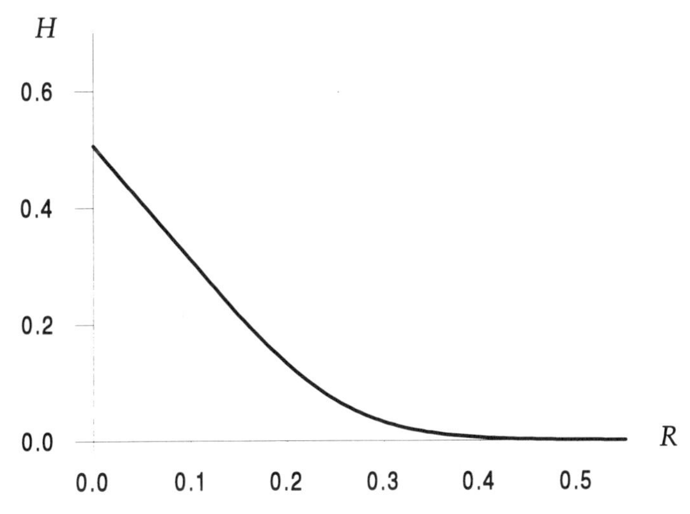

Figure 14.3 The European average strike call option: H versus R with $\sigma = 0.4$ and $r = 0.1$ at three months before expiry; there has already been three months' averaging.

applying numerical methods directly to the partial differential equation. In Figure 14.3 we see H against R at three months before expiry and with three months' averaging completed; $\sigma = 0.4$ and $r = 0.1$.

In the case of an American option, we have to solve the partial differential inequality (14.7) subject to the constraint (14.8), the final condition and the condition that $H \to 0$ as $R \to \infty$. We cannot do this analytically and we must find the solution numerically.

14.4.1 Put-call Parity for the European Average Strike

The payoff at expiry for a portfolio of one European average strike call held long and one put held short is

$$S \max(1 - R/T, 0) - S \max(R/T - 1, 0).$$

Whether R is greater or less than T at expiry, this payoff is simply

$$S - \frac{RS}{T}.$$

The value of this portfolio is identical to one consisting of one asset and a financial product whose payoff is

$$-\frac{RS}{T}.$$

In order to value this product we seek a solution of the average strike equation of the form

$$H(R,t) = a(t) + b(t)R \tag{14.10}$$

and with $a(T) = 0$ and $b(T) = -1/T$; such a solution would have the required payoff of $-RS/T$. Substituting (14.10) into (14.7) and satisfying the boundary conditions, we find that

$$a(t) = -\frac{1}{rT}\left(1 - e^{-r(T-t)}\right), \quad b(t) = -\frac{1}{T}e^{-r(T-t)}.$$

We conclude that

$$C - P = S - \frac{S}{rT}\left(1 - e^{-r(T-t)}\right) - \frac{1}{T}e^{-r(T-t)}\int_0^t S(\tau)\,d\tau,$$

where C and P are the values of the European arithmetic average strike call and put. This is put-call parity for the European average strike option.

14.5 Average Rate Options

The typical average rate option is very similar to the average strike option in that the payoff depends on a suitably defined average of the asset price. The difference is in the structural form of the payoff. Whereas the average strike is the same as a vanilla option except that the exercise price is replaced by the average, the average rate has the same payoff as the vanilla but with the asset price replaced by the average. That is, an arithmetic average rate call option has payoff given by

$$\max(I/T - E, 0)$$

at expiry. Such options are usually more difficult to value than average strike options because, as a rule, they do not admit similarity reductions: typically, it is not possible to reduce the number of independent variables from three to two. Generally, therefore, such problems must be solved numerically. However, when the average is measured geometrically, it is possible to reduce the problem to one in two variables and we now show how to find explicit formulæ.

14.5.1 Geometric Averaging and Continuous Sampling

There are explicit solutions for the value of average rate options when the average is measured geometrically. This is because the logarithm of the asset price follows a random walk with variance which is independent of the asset price. There is also an intimate link between sums of logarithms and the geometric average.

The explicit formulæ exist only for European options. Let us therefore consider a European average rate option with payoff at expiry given by

$$V(S, I, T) = \Lambda(I).$$

Observe that the payoff is here only a function of I and not of S; this makes an explicit solution possible.

When the geometric average is sampled continuously, I is given by

$$I = \int_0^t \log S(\tau)\, d\tau$$

and, recalling the analysis in Section 13.2, for a European option we must solve

$$\frac{\partial V}{\partial t} + \log S \frac{\partial V}{\partial I} + \tfrac{1}{2}\sigma^2 S^2 \frac{\partial^2 V}{\partial S^2} + rS\frac{\partial V}{\partial S} - rV = 0. \tag{14.11}$$

If the payoff is a function of I only, we can seek a solution of the form $F(y, t)$, where now

$$y = \frac{I + (T - t)\log S}{T}.$$

With this independent variable the differential equation becomes a parabolic partial differential equation with coefficients that are independent of y and, moreover, the log term is eliminated:

$$\frac{\partial F}{\partial t} + \frac{1}{2}\left(\frac{\sigma(T-t)}{T}\right)^2 \frac{\partial^2 F}{\partial y^2} + (r - \tfrac{1}{2}\sigma^2)\left(\frac{T-t}{T}\right)\frac{\partial F}{\partial y} - rF = 0. \tag{14.12}$$

(We leave the derivation of this equation as an exercise.)

The coefficients of (14.12) are independent of y but are, however, functions of time, t. Therefore, what remains after the log terms are removed from (14.11) is almost the Black–Scholes equation, under a logarithmic transformation with time-dependent volatility and interest rate and with a nonzero, time-dependent dividend yield.

In Section 6.5 we saw that the Black–Scholes formulæ need only very simple modifications to yield explicit formulæ for time varying volatility, interest rate and dividend yield. For options with more complicated payoffs than calls or puts it is still possible to solve the constant coefficient partial differential equation by taking advantage of the general solution of the Black–Scholes equation with arbitrary final condition as given in equation (5.16) of Section 5.5.

Explicit formulæ may be found, and the following rules, the derivation of which is left as an exercise, show how to convert an explicit Black–Scholes formula for a vanilla option to an explicit formula for a geometric average rate option:

- Take the Black–Scholes formula for a vanilla option having the same payoff as the Asian, but in terms of S instead of $e^{I/T}$; for example

$$\max(S - E, 0) \quad \text{instead of} \quad \max(e^{I/T} - E, 0).$$

  Call this $V_{BS}(S, t; r, \sigma)$. (In the example above, $V_{BS}(S, t; r, \sigma)$ is the formula for an option with payoff $\max(S - E, 0)$, i.e. a European call.)
- Wherever σ^2 appears in the formula for V_{BS}, replace it by

$$\frac{1}{T-t}\int_t^T \sigma^2 \frac{(T-\tau)^2}{T^2}\, d\tau.$$

  Thus, if σ is constant, the effective volatility is $\sigma^2 (T-t)^2/3T^2$.
- Wherever r appears in the formula, replace it by

$$\frac{1}{T-t}\int_t^T \left(r - \tfrac{1}{2}\sigma^2\right)\frac{(T-\tau)}{T}\, d\tau.$$

  When r and σ are constant, the effective interest rate is thus $(\frac{1}{2}\sigma^2 - r)(T-t)/2T$.
- Multiply the resulting formula by

$$\exp\left(-\int_t^T \left(r - \frac{(T-\tau)}{T}\left(r - \tfrac{1}{2}\sigma^2\right)\right) d\tau\right).$$

  When r and σ are constant this factor is

$$\exp\Big(-\tfrac{1}{2}\big(\sigma^2 + (r - \tfrac{1}{2}\sigma^2)(T+t)\big)(T-t)\Big).$$

- Replace S by $e^{I/T} S^{(T-t)/T}$.

The above discussion centred on the solution of the European continuously sampled average rate option. The same idea can be applied to the discrete sampling case, when again explicit formulæ can be derived. We leave this as an exercise for the reader.

14.6 Discretely Sampled Averages

In practice it can be difficult to calculate the average of an asset price from its complete time series: prices can change every 30 seconds or so, with the occasional misquotation of prices. Thus, it is more common in an option contract to specify that the average is to be calculated from a small subset of the complete time series for the asset price, for example the average over the daily or weekly closing prices.

We have already modelled the continuous average as an integral, consistently with the assumption that asset and option values are time-continuous quantities. By a discrete average we mean the sum, rather than the integral, of a finite number of values of the asset during the life of the option. Such a definition of average is easily included within the framework of our model. The discrete sampling of averages bears close similarities with the discrete payment of dividends. In particular, both give rise to jump conditions across payment/sampling dates. These ideas have all been discussed before in Chapter 13; here we recall the basic methodology and results.

We first recall the jump condition for an Asian option with discrete arithmetic averaging; this derivation is easily generalised to Asian options with more complicated discrete averaging, for example discrete geometric averaging.

The discretely sampled arithmetic running sum may be defined as

$$I = \sum_{i=1}^{j(t)} S(t_i),$$

where t_i are the sampling dates and $j(t)$ is the largest integer such that $t_{j(t)} < t$. In terms of I and $j(t)$, the discretely sampled arithmetic average is $I/j(t)$.

Across a sampling date this running sum is necessarily discontinuous. It is updated from a value I before the sampling to the new value given by $I + S$. Since the realised option value must be continuous across the sampling date we find that for the discretely sampled arithmetic average strike option,

$$V(S, I, t_i^-) = V(S, I + S, t_i^+). \tag{14.13}$$

Likewise, for the discretely sampled geometric average where the running sum is

$$I = \sum_{i=1}^{j(t)} \log S(t_i)$$

the jump condition is

$$V(S, I, t_i^-) = V(S, I + \log S, t_i^+).$$

Because the path-dependent quantity, I, is updated discretely and is therefore constant between sampling dates, the partial differential equation for the option value between sampling dates is just the basic Black–Scholes equation with I treated as a parameter. Thus the strategy for valuing any Asian option is as follows:

- Working backwards from expiry, solve

$$\frac{\partial V}{\partial t} + \tfrac{1}{2}\sigma^2 S^2 \frac{\partial^2 V}{\partial S^2} + rS\frac{\partial V}{\partial S} - rV = 0$$

  between sampling dates.
- Then apply the appropriate jump condition across the current sampling date to deduce the option value immediately before the present sampling date.
- Repeat this process as necessary to arrive at the current value of the option.

Further Reading

- More details about the partial differential equation approach to the valuation of Asian options can be found in *Option Pricing.*
- Some exact solutions can be found in Boyle (1991) and, for geometric averaging, in Rubinstein (1992).
- Ingersoll (1987) presents the partial differential equation formulation of some average strike options and demonstrates the similarity reduction.
- For other methods of evaluating Asian options see Geman & Yor (1992).
- The application of the numerical Monte Carlo method is described by Kemna & Vorst (1990).
- More examples of the methods described here can be found in Dewynne & Wilmott (1993 c, d).
- For an approximate valuation of arithmetic Asian options see Levy (1990), who replaces the density function for the distribution of the average by a lognormal function.

Exercises

1. Find formulæ for the value of some perpetual options, i.e. having no time dependence. These are very simple similarity solutions.

2. Repeat the analysis of the European continuously sampled geometric average rate option when the average is measured discretely. Derive explicit formulæ for the value of such options.

3. We have seen the equation satisfied by the European continuously sampled arithmetic average rate call option. In special circumstances it can be known with certainty before expiry that the option will expire in the money. What are these circumstances? Find an explicit formula for the value of the option in this case.

   Extend this idea to other types of Asian option, and derive further formulæ.

4. What is the partial differential equation for the value of an option that depends on

$$\left(\int_0^T \Big(S(\tau) \Big)^n d\tau \right)^{1/n} ?$$

   Consider this Asian option with different payoffs. Determine whether similarity reductions are possible for your examples. Look for simple explicit formulæ.

5. The average strike foreign exchange option has the payoff

$$\max\left(1 - \frac{1}{ST} \int_0^T S(\tau)\, d\tau, 0 \right),$$

   where S is an exchange rate. What is the partial differential equation satisfied by this option? (Remember the interest payments on the foreign currency.) Is there a similarity reduction?

6. Recall the jump conditions for the discretely sampled arithmetic average strike option. In this case the option price has a similarity reduction of the form $V(S, I, t) = SH(I/S, t)$. Write the jump conditions in terms of $H(R, t)$, where $R = I/S$.

7. Find similarity variables for the discretely and continuously sampled *geometric* Asian options. What form must the payoff function take?

15 Lookback Options: Options on the Maximim or Minimum

15.1 Introduction

A lookback option is a derivative product whose payoff depends on the maximum or minimum realised asset price over the life of the option. For example, a **lookback put** has a payoff at expiry that is the difference between the maximum realised price and the spot price at expiry. This may be written as

$$\max(J - S, 0)$$

where J is the maximum realised price of the asset:

$$J = \max_{0 \leq \tau \leq t} S(\tau).$$

As for the Asian options considered in the previous chapter, the maximum or minimum realised asset price may be measured continuously or, more commonly, discretely.

Such options give the holder an extremely advantageous payoff. Using lookback options, one can construct a product enabling the investor to buy at the low and sell at the high. They are therefore relatively expensive.

We continue in the spirit of the analysis in Chapters 13 and 14. The general framework established in those chapters is sufficiently robust to include both European and American exercise features and continuous and discrete sampling; we now apply this framework to lookback options. As in the previous chapter we find that European lookbacks lead to partial differential equations with final and boundary conditions, whereas for American lookbacks we obtain a partial differential inequality subject to a constraint and, consequently, a linear complementarity problem. With discrete sampling we find that, as before, jump conditions apply across sampling dates. In general the option value is a function of the three variables S, J and t, but we also find that if the payoff has a certain form then the problem admits a similarity reduction to two independent variables. When a similarity reduction can be found the option value may, in some cases, be determined explicitly. Even if the solution cannot be found explicitly, the similarity reduction allows more efficient numerical solution.

As in the examples in Chapter 14 on Asian options we consider, amongst others, the lookback *strike* and the lookback *rate*, in both call and put varieties. If J is the sampled maximum, the **lookback strike** put option has a payoff similar to a vanilla put but with J replacing the exercise price E, i.e. the payoff is

$$V(S, J, T) = \max(J - S, 0).$$

(This option admits a similarity reduction in the variables S/J and t.) Similarly, the **lookback rate** put has payoff similar to the vanilla put but with J replacing S, i.e.

$$V(S, J, T) = \max(E - J, 0),$$

where E is prescribed. (This option does not admit a similarity reduction and must be solved in three dimensions.)

We continue to work in the general framework introduced in Chapter 13 and then consider some special cases. At the end of the chapter we consider two perpetual options which depend on the maximum realised asset price, the 'Russian' and the 'stop-loss'. Both of these have simple exact solutions.

We concentrate on valuing a put option. The equivalent call option depends on the realised minimum of the asset price

$$J = \min_{0 \leq \tau \leq t} S(\tau)$$

but is otherwise similar to the put option, and we leave the details of its valuation as an exercise.

15.2 Continuous Sampling of the Maximum

In this section we consider a put option that depends on the maximum value of the asset where the maximum is measured continuously, as illustrated in Figure 15.1. Observe that, when the maximum is updated

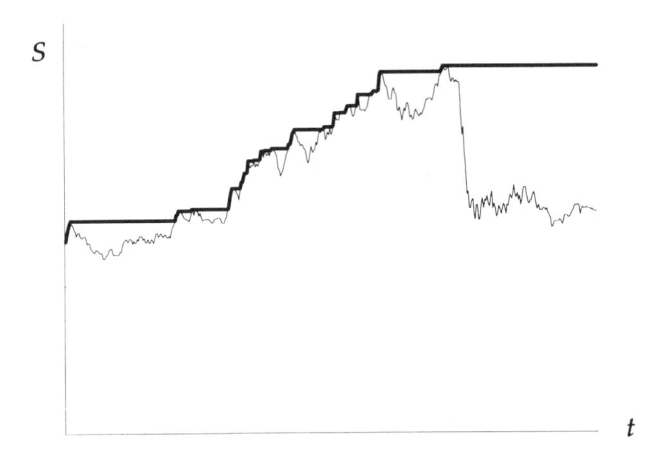

Figure 15.1. An example of continuous measurement of the maximum.

continuously, the asset price is necessarily less than or equal to the maximum:

$$0 \leq S \leq J.$$

Since the lookback put is a path-dependent option, its value P is not simply a function of S and t, as is the case for a simple option. If the independent variable J is the maximum realised asset value over the life of the option, P also depends on J since it depends on J at expiry. Thus

$$P = P(S, J, t).$$

It may not be immediately obvious how our general exotic option framework, for which the path-dependent quantity has been defined as an integral, can accommodate the lookback case. This is not as difficult as might be supposed. Let us define

$$I_n = \int_0^t (S(\tau))^n \, d\tau. \tag{15.1}$$

Now if we introduce

$$J_n = (I_n)^{1/n}$$

this places the lookback option in our general setting. Why is this so? The strategy is to consider an option whose value depends on J_n and then take the limit as $n \to \infty$. As n tends to infinity, we have[1]

$$J = \lim_{n\to\infty} J_n = \max_{0\le\tau\le t} S(\tau),$$

and the relevance to lookback options is clear. (Similarly as $n \to -\infty$, $J_n \to \min_{0\le\tau\le t} S(\tau)$.)

Now we derive the stochastic differential equation satisfied by J_n; the argument exactly parallels that given in Chapter 13 for the general case. In the time t to $t + dt$, J_n changes by an amount dJ_n given by

$$J_n + dJ_n = \left(\int_0^{t+dt} (S(\tau))^n \, d\tau\right)^{1/n}.$$

From this and (2.1) we see that

$$dJ_n = \frac{1}{n}\frac{S^n}{(J_n)^{n-1}}\, dt. \tag{15.2}$$

Thus J_n is a deterministic variable, as there are no random terms in (15.2). We need (15.2) to apply Itô's lemma to P.

As we have done many times before we construct a hedged portfolio consisting of one option and a number $-\Delta$ of the underlying asset:

$$\Pi = P - \Delta S.$$

In the time from t to $t + dt$ the value of this portfolio changes by an amount $d\Pi$ given by

$$d\Pi = dP - \Delta\, dS.$$

Choosing

$$\Delta = \frac{\partial P}{\partial S}$$

and using Itô's lemma to expand dP, remembering that P depends on the three variables S, J_n and t, we find that

$$d\Pi = \frac{\partial P}{\partial t} dt + \frac{1}{n}\frac{S^n}{(J_n)^{n-1}}\frac{\partial P}{\partial J_n} dt + \tfrac{1}{2}\sigma^2 S^2 \frac{\partial^2 P}{\partial S^2} dt. \tag{15.3}$$

[1] At this point it is important to note that $S(\tau)$ is a *continuous* realisation of the random walk (2.1). If $S(\tau)$ in this integral is not continuous the result need not follow. We have not previously had cause to discuss the continuity of realisations of (2.1), and it is not obvious that we can make the assumption that $S(\tau)$ is continuous. A full discussion of the matter of continuous realisations is outside the scope of this text, and we refer the reader to a text on stochastic calculus (for example, one of those cited in Chapter 2). The basic result is that we can assume the continuity without loss of generality.

When the option is American there may be times when it is optimal to exercise the option before expiry. We can only insist that the return is at most that to be received from a risk-free account, thus

$$d\Pi \le r\Pi\, dt = r\left(P - S\frac{\partial P}{\partial S}\right) dt. \tag{15.4}$$

In the case of a European option we have equality in (15.4). Bringing together (15.3) and (15.4) we arrive at

$$\frac{\partial P}{\partial t} + \frac{1}{n}\frac{S^n}{(J_n)^{n-1}}\frac{\partial P}{\partial J_n} + \tfrac{1}{2}\sigma^2 S^2 \frac{\partial^2 P}{\partial S^2} + rS\frac{\partial P}{\partial S} - rP \le 0. \tag{15.5}$$

We now take the limit $n \to \infty$. Since $S \le \max S = J$, in this limit the coefficient of $\partial V/\partial J_n$ tends to zero. Thus in this limit the partial differential inequality becomes

$$\frac{\partial P}{\partial t} + \tfrac{1}{2}\sigma^2 S^2 \frac{\partial^2 P}{\partial S^2} + rS\frac{\partial P}{\partial S} - rP \le 0. \tag{15.6}$$

This is simply the usual Black–Scholes inequality; for a European option it is the Black–Scholes equation. The independent variable J appears as a parameter only in this equation, but it also features in the boundary and final conditions.

The final condition for the equation is simply the payoff at expiry. The lookback put has

$$P(S, J, T) = \max(J - S, 0). \tag{15.7}$$

This is the final condition regardless of whether the option is European or American or whether the sampling is continuous or discrete.

When the maximum is sampled continuously it is impossible for the asset price ever to exceed the sampled maximum; thus $S \le J$. Therefore the problem is posed only on the region $0 \le S \le J$.

It is interesting to note that because $S \le J$, the lookback put with a continuously sampled maximum is, in a sense, not an option: with probability one, it will be exercised. (The only case where it may not be exercised is in the unlikely event that the maximum realised asset value occurs at expiry.)

15.2.1 The European Case

When the option is European, arbitragers can hold both sides of the Black–Scholes portfolio, and so the inequality in (15.6) becomes equality.

We have a final condition from (15.7), and boundary conditions are applied at $S = 0$ and $S = J$.

If S is zero then it can never become greater than zero. The payoff at time T is known with certainty to be J. Hence the interest-rate discounted present value of the option is

$$P(0, J, t) = Je^{-r(T-t)}. \tag{15.8}$$

The remaining boundary condition comes from considering the behaviour of the random walk close to the boundary $S = J$. Suppose that, at some time prior to expiry, S is close to its maximum realised so far, i.e. S is close to J. It can be shown that the probability that the current value of the maximum is still the maximum at expiry is zero. Since the present value of the maximum is not the final maximum, the value of the option must be insensitive to small changes in J. The remaining boundary condition is therefore

$$\frac{\partial P}{\partial J} = 0 \quad \text{on} \quad S = J. \tag{15.9}$$

These final and boundary conditions give a unique value for the option.

The solution for the lookback put is

$$S\left(-1 + N(d_7)(1 + k^{-1})\right) + Je^{-r(T-t)}\left(N(d_5) - k^{-1}\left(\frac{S}{J}\right)^{1-k} N(d_6)\right),$$

where

$$d_5 = \frac{\log(J/S) - (r - \frac{1}{2}\sigma^2)T - t}{\sigma\sqrt{T-t}},$$

$$d_6 = \frac{\log(S/J) - (r - \frac{1}{2}\sigma^2)T - t}{\sigma\sqrt{T-t}},$$

$$d_7 = \frac{\log(S/J) + (r + \frac{1}{2}\sigma^2)T - t}{\sigma\sqrt{T-t}}$$

and

$$k = r/\tfrac{1}{2}\sigma^2.$$

This formula can be derived by an extension of the method of images, and we hint at how this can be done later in this chapter.

15.2.2 The American Lookback Put

When early exercise is a possibility, the exercise price of the option must be specified for times prior to expiry. The natural specification for a lookback put is

$$\Lambda(S, J, t) = \max(J - S, 0).$$

When the option is American the possibility of early exercise means that we have inequality in (15.6). There may be times at which it is optimal to exercise the option as well as times when it should be held. For the lookback put we see that if S is less than some critical value $S_f(J,t)$ then it is, in fact, optimal to exercise.

Arbitrage considerations show that the value of the option must satisfy the constraint

$$P(S, J, t) \geq \Lambda(S, J, t), \tag{15.10}$$

since this is the payoff for early exercise. Further, all of P, $\partial P/\partial S$ and $\partial P/\partial J$ must be continuous.

The final condition (15.7) is satisfied at $t = T$ and, should the boundaries at $S = 0$ and/or $S = J$ ever lie in the hold region, the boundary conditions at $S = 0$, (15.8), and/or at $S = J$, (15.9), must also be satisfied.

For times before expiry $t < T$, $S = 0$ cannot lie in the hold region. This follows from (15.8) and (15.10); if $S = 0$ does lie in the hold region then

$$P(0, J, t) = Je^{-r(T-t)} < \Lambda(0, J, t) = J,$$

contrary to (15.10). Thus there must be an optimal exercise boundary $S_f(J,t)$ separating an early exercise region where $S < S_f(J,t)$ from a hold region where $S > S_f(J,t)$.

We can write this problem in linear complementarity form (and thereby eliminate explicit reference to $S_f(J,t)$) as follows. Define the linear operator $\mathcal{L}_{BS}$ by

$$\mathcal{L}_{BS}(\cdot) = \frac{\partial}{\partial t} + \tfrac{1}{2}\sigma^2 S^2 \frac{\partial^2}{\partial S^2} + rS\frac{\partial}{\partial S} - r;$$

this is, of course, the basic Black–Scholes operator. The American lookback put problem can be written as

$$\mathcal{L}_{BS}(P) \leq 0 \quad \text{and} \quad \Big(P - \Lambda(S, J, t)\Big) \geq 0, \tag{15.11}$$

together with

$$\mathcal{L}_{BS}(P) \cdot (P - \Lambda) = 0, \tag{15.12}$$

and the final condition

$$P(S, J, T) = \Lambda(S, J, T), \tag{15.13}$$

with the boundary condition

$$\frac{\partial P}{\partial J}(J, J) = 0. \tag{15.14}$$

The problem is to be solved only for $0 \leq S \leq J$; all of P, $\partial P/\partial S$ and $\partial P/\partial J$ must be continuous.

15.3 Discrete Sampling of the Maximum

Most commercial lookback contracts are based on a discretely measured maximum or minimum. This is for two reasons. The first reason is contractual; it is easier to measure the maximum of a small set of values, all of which can be guaranteed to be 'real' prices at which the underlying has traded. The second reason is that by decreasing the frequency at which the maximum is measured, some contracts become cheaper and therefore more appealing. Figure 15.2 shows an example of discrete measurement of the maximum. The ticks on the horizontal time axis represent the times at which the maximum is sampled. As we can see, the asset price can now exceed the sampled maximum if it does so between sampling dates. With J still denoting the maximum, albeit sampled discretely, it is no longer true that S must always be less than J. This is a very important difference between the continuously and discretely sampled cases: the domain on which the problem is posed is quite different. The payoff for the lookback put is still $\max(J - S, 0)$ but there now arises the possibility that the option will not be exercised; it will not be exercised at expiry if $S > J$.

When the maximum is sampled discretely we still obtain the Black–Scholes equation or inequality in S and t, with J entering only as a parameter. Across the sampling dates there is a jump condition. The financial argument for the jump condition is similar to that used in Chapter 6 for discrete dividends and in Chapters 13 and 14. Arbitrage considerations show that the realised value of the option cannot be discontinuous. Thus $P(S, J, t)$ must be continuous as S, J and t vary along any given realisation. Across a sampling date the discretely sampled maximum is updated according to the rule

$$J_i = \max(J_{i-1}, S),$$

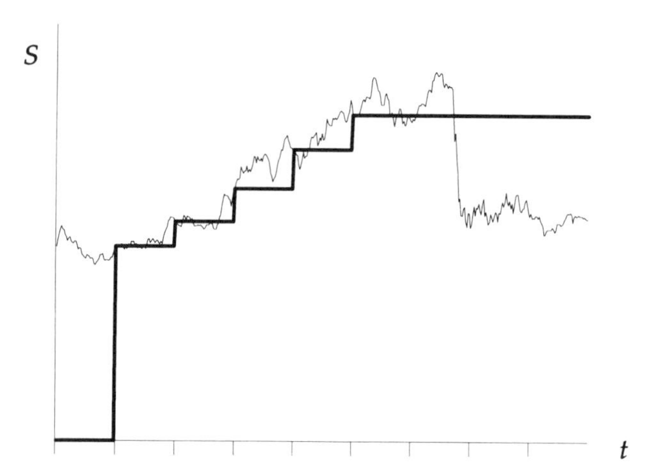

Figure 15.2 A schematic diagram of the discretely measured maximum. The ticks on the time axis are the sampling dates; J does not change between these dates.

where J_i is the value of J just after sampling at time t_i. Note that J then remains constant until immediately after t_{i+1}. Continuity of the realised put option price may be stated as

$$P(S_i, J_{i-1}, t_i^-) = P(S_i, J_i, t_i^+),$$

where as before S_i is the value of S at time t_i. Writing J for $J(t_{i-1})$, this is equivalent to the jump condition

$$P(S, J, t_i^-) = P\left(S, \max(J, S), t_i^+\right)$$

in a Black–Scholes framework where S and J are considered independent variables.

15.4 Similarity Reductions

The formulation of the problem has so far used the 'primitive variables' S, J, t. In this section we show how to recast some lookback problems in terms of only two variables; we thus seek similarity reductions and solutions. The lookback option model presented so far is very general in that it permits an arbitrary payoff (at expiry or earlier) $\Lambda(S, J, t)$.

If we make some restrictions on the class of payoff structures that are admissible, we may take the analysis further and find classes of lookbacks that depend on only time and a single state variable. The restriction that we make is

- the payoff has the form $\Lambda(S, J, t) = J\hat{\Lambda}(S/J, t)$.

The lookback put has payoff

$$J \max(1 - S/J, 0)$$

and therefore satisfies this requirement.

With this restriction we may find a solution of the form

$$P(S, J, t) = JW(\xi, t),$$

where

$$\xi = S/J.$$

In the following we describe only the European option; the modification for an American option is simple and left as an exercise.

With these definitions for W and ξ we find that the partial differential equation for $W(\xi, t)$ is

$$\frac{\partial W}{\partial t} + \tfrac{1}{2}\sigma^2\xi^2\frac{\partial^2 W}{\partial \xi^2} + r\xi\frac{\partial W}{\partial \xi} - rW = 0$$

and the boundary condition at $S = 0$ becomes

$$W(0, t) = e^{-r(T-t)}.$$

The final condition for a lookback put becomes

$$W(\xi, T) = \max(1 - \xi, 0).$$

When the maximum is measured continuously, the boundary condition at $S = J$ becomes a boundary condition at $\xi = 1$ and is

$$\frac{\partial W}{\partial \xi} = W \quad \text{on} \quad \xi = 1.$$

If the maximum is measured discretely, the boundary condition as $S \to \infty$ becomes a boundary condition as $\xi \to \infty$, and is

$$\frac{\xi}{W}\frac{\partial W}{\partial \xi} \sim 1 \quad \text{as} \quad \xi \to \infty.$$

The jump condition across sampling dates becomes

$$W(\xi, t_i^-) = \max(\xi, 1)\, W\left(\min(\xi, 1), t_i^+\right).$$

ξ	A	B	C	O
0.9	0.125	0.120	0.114	0.104
1.0	0.105	0.095	0.081	0.048
1.1	0.111	0.098	0.082	0.021

Figure 15.3 American lookback put values for $r = 0.1$, $\sigma = 0.2$, $T = 1.0$. Cases A, B and C correspond to sampling at different times (see text for details). Case O corresponds to no sampling; this is the simple put with exercise price 1.0.

Recall the exercise at the end of Chapter 12 in which it was demonstrated that if $V(S,t)$ satisfies the Black–Scholes equation then $U(S,t) = S^{\alpha}V(a/S,t)$ satisfies a similar partial differential equation. In particular, for the choice

$$\alpha = 1 - r/\tfrac{1}{2}\sigma^2$$

it is easily shown that U satisfies the Black–Scholes equation. This observation, together with the above similarity reduction, make it relatively straightforward to find the explicit formulæ for the lookback put. This is left as an exercise for the reader.

15.5 Some Numerical Examples

We do not describe numerical methods for exotic options in this book; they are covered in *Option Pricing.* For the benefit of readers who would like to extend the methods we do describe, in this section we give some numerical results for simple lookback puts. The numerical methods used to generate them are described in *Option Pricing.*

In Figures 15.3 and 15.4 we see comparisons between American and European option prices using different sets of sampling times. In each case the values shown are at one year before expiry with zero dividend yield, $r = 0.1$ and $\sigma = 0.2$. Figure 15.3 shows option values for the American option and Figure 15.4 for the equivalent European option.

There are three examples of the sampling of the maximum and one simple vanilla put problem for comparison. The latter is Case O; this

ξ	A	B	C	O
0.9	0.101	0.094	0.087	0.074
1.0	0.089	0.079	0.067	0.038
1.1	0.095	0.083	0.068	0.017

Figure 15.4 European lookback put values for $r = 0.1$, $\sigma = 0.2$, $T = 1.0$. Cases A, B and C correspond to sampling at different times (see text for details). Case O corresponds to no sampling; this is the simple put with exercise price 1.0.

is exactly equivalent to a vanilla put option with unit exercise price. Otherwise we have

Case A: sampling at times 0.5, 1.5, 2.5, ... 10.5, 11.5 months;

Case B: sampling at times 1.5, 3.5, 5.5, 7.5, 9.5, 11.5 months;

Case C: sampling at times 3.5, 7.5, 11.5 months.

The tables should be read as follows. Suppose the value of an American lookback put option with discrete sampling under sampling strategy B is required. Recall that the value of the option is given by

$$P = JW(S/J, t).$$

If we want the value of the option at one year to expiry when the asset price is 180 and the current maximum is 200 then we must look along the row $\xi = S/J = 180/200 = 0.9$. The value of the option is then $200 \times 0.120 = 24$.

Observe that the option price decreases as the number of samples decreases (from A to C). This is financially obvious, since the fewer samples the lower the final payoff is likely to be. Decreasing the frequency of measurement of the maximum decreases their cost. This may be important, since one of the commercial criticisms of lookback options is that they are too expensive. Also note that the option price reaches a minimum close to $\xi = 1$. The option delta can become positive, since it is beneficial for the holder of the option if the asset price rises just before a sampling date and then falls. As expected, American prices are everywhere greater than European.

15.6 Two 'Perpetual Options'

We have seen one explicit formula for a simple lookback put, and in the exercise at the end of this chapter we make suggestions for finding more. In this section we find two more explicit formulæ, for a 'Russian' option and a 'stop-loss' option. Both of these lookback options share the property that they are perpetual options, i.e. they do not have an expiry date but rather an infinite time horizon. (Another perpetual option was the perpetual barrier option, an exercise in Chapter 10.)

15.6.1 Russian Options

A **Russian option** is a perpetual American option, which, at any time chosen by the holder, pays out the maximum realised asset price up to that date. We consider only the continuously sampled maximum case, as it is unlikely that a tractable explicit solution exists if the sampling is discrete. To make the problem interesting we assume that there is a continuously paid constant dividend yield, as discussed in Chapter 6; without dividends the problem is trivial.

As the time horizon is infinite we may take the option value to be *independent of time:* $V = V(S, J)$. (With discrete sampling at periodic intervals the solution would be periodic.) As before, let J be the maximum realised value of the asset price. When it is optimal to hold and the option exists, we solve the time-independent Black–Scholes equation

$$\tfrac{1}{2}\sigma^2 S^2 \frac{\partial^2 V}{\partial S^2} + (r - D_0) S \frac{\partial V}{\partial S} - rV = 0$$

(note the dividend term) with the boundary condition

$$\frac{\partial V}{\partial J} = 0 \quad \text{on} \quad J = S.$$

The solution must also satisfy

$$V \geq J,$$

since the option is American and the right-hand side of this inequality is the early exercise payoff. There must be a free boundary, since this option is pointless if it is never exercised, and both V and $\partial V/\partial S$ must be continuous there.

Let us seek a solution in the form

$$V = JW(\xi)$$

where $\xi = S/J$. Then we have

$$\tfrac{1}{2}\sigma^2\xi^2 W'' + (r - D_0)\xi W' - rW = 0, \tag{15.15}$$

where $'$ denotes $d/d\xi$. Suppose that the free boundary is at $\xi = \xi_0$. The boundary conditions become

$$W - W' = 0 \quad \text{at} \quad \xi = 1$$

and

$$W = 1 \quad \text{and} \quad W' = 0 \quad \text{at} \quad \xi = \xi_0.$$

The general solution of (15.15) is found by trying $W = \text{constant} \times \xi^\alpha$ for constant α. This yields a quadratic equation for α, whose roots are

$$\alpha_\pm = \frac{1}{\sigma^2}\left(-r + D_0 + \tfrac{1}{2}\sigma^2 \pm \sqrt{(r - D_0 - \tfrac{1}{2}\sigma^2)^2 + 2\sigma^2 r} \right). \tag{15.16}$$

The solution to the boundary value problem is then easily found to be

$$W = \frac{1}{\alpha_+ - \alpha_-}\left(\alpha_+ \left(\frac{\xi}{\xi_0}\right)^{\alpha_-} - \alpha_- \left(\frac{\xi}{\xi_0}\right)^{\alpha_+} \right),$$

where the free boundary conditions give

$$\xi_0 = \left(\frac{\alpha_+(1 - \alpha_-)}{\alpha_-(1 - \alpha_+)} \right)^{1/(\alpha_- - \alpha_+)}.$$

When the dividend yield is zero, i.e. $D_0 = 0$, the problem does not have a solution. It is, clearly, never optimal to hold such an option when the underlying does not pay dividends.

15.6.2 The Stop-loss Option

A **stop-loss option** may be thought of as a perpetual barrier lookback with a rebate that is a fixed proportion of the maximum realised value of the asset price. Thus, if S reaches a maximum value J and then falls back to λJ, where $\lambda < 1$, the option pays the owner S (which at that time is equal to λJ). It has an obvious use to lock in a good proportion of a profit while relieving the owner of the uncertainty of guessing when the maximum is reached. Note that the option is not triggered until the fall occurs.

Since the option pays the owner the amount S when S reaches λJ, we have

$$V(\lambda J, J) = \lambda J. \tag{15.17}$$

We again write $V = JW(\xi)$ where, as before, $\xi = S/J$. The differential equation is again (15.15), while (15.17) becomes

$$W(\lambda) = \lambda,$$

and the remaining boundary condition is

$$W - W' = 0 \quad \text{at} \quad \xi = 1.$$

The solution is

$$W = \lambda \frac{\xi^{\alpha_+}(1-\alpha_-) - \xi^{\alpha_-}(1-\alpha_+)}{\lambda^{\alpha_+}(1-\alpha_-) - \lambda^{\alpha_-}(1-\alpha_+)},$$

where $\alpha_\pm$ are given by (15.16). When $D_0 = 0$ the solution is

$$W = \xi$$

irrespective of λ, i.e. $V = S$: the option is equivalent to the underlying.

Further Reading

- More details about the valuation of lookback options can be found in *Option Pricing.*
- Lookback options were first described in the academic literature by Goldman, Sosin & Gatto (1979), who presented an explicit formula for a European option where the maximum is measured continuously throughout the life of the option.
- Some more explicit solutions are given by Conze & Viswanathan (1991).
- More numerical results can be found in Dewynne & Wilmott (1993b).
- Russian options are described by Duffie & Harrison (1992).
- The stop-loss option when the maximum is measured discretely is covered in Fitt, Wilmott & Dewynne (1994).

Exercises

1. Derive the explicit formula for the value of a European lookback put.
2. Find explicit formulæ in the following cases, all with continuous sampling of the maximum or minimum:
   (a) lookback call, with payoff $\max(S - J, 0)$, where J is the asset price minimum;

   (b) lookback calls and puts with constant dividend yield;
   (c) lookback calls and puts with time varying volatility, interest rate and dividend yield.

3. There is no reason why sampling dates must be evenly spaced. How do you expect lookback option prices to be affected by the structure of the sampling?

16 Options with Transaction Costs

16.1 Introduction

We have derived the Black–Scholes partial differential equation for simple option prices, we have discussed the general theory behind the diffusion equation and, in the last few chapters, we have generalised the Black–Scholes model to exotic options. We continue this generalisation with a model that incorporates the effects of transaction costs on a hedged portfolio. We describe the model only for vanilla options but it can easily be modified for exotic options.

16.2 Discrete Hedging

One of the key assumptions of the Black–Scholes analysis is that the portfolio is rehedged continuously: we take the limit $dt \to 0$. If the costs associated with rehedging (e.g. bid-offer spread on the underlying) are independent of the timescale of rehedging then the infinite number of transactions needed to maintain a hedged position until expiry may lead to infinite total transaction costs. Since the Black–Scholes analysis is based on a hedged portfolio, the consequences of significant costs associated with rehedging are important. Different people have different levels of transaction costs; as a general rule there are economies of scale, so that the larger the trader's book, the less significant are his costs. Thus, contrary to the basic Black–Scholes model, we may expect that there is no unique option value. Instead, the value of the option depends on the investor.

Leland has proposed a very simple modification to the Black–Scholes model for vanilla calls and puts, which can be extended to portfolios of options, that introduces discrete revision of the portfolio and transaction costs. In the main his assumptions are those mentioned in Chapter 3 for the Black–Scholes model with the following exceptions:

- The portfolio is revised every δt where now δt is a non-infinitesimal fixed time-step; note that we do not take $\delta t \to 0$. For example, the portfolio may be rehedged every day at 9:00 a.m.
- The random walk is given in discrete time by

$$\delta S = \sigma S \phi \sqrt{\delta t} + \mu S \, \delta t$$

  where ϕ is drawn from a standardised normal distribution.
- Transaction costs in buying or selling the asset are proportional to the monetary value of the transaction. Thus if ν shares are bought ($\nu > 0$) or sold ($\nu < 0$) at a price S, then the transaction costs are $\kappa|\nu|S$, where κ is a constant depending on the individual investor. A more complex cost structure can be incorporated into the model with only a small amount of effort (see the Exercises at the end of this chapter).
- The hedged portfolio has an *expected* return equal to that from a bank deposit.

We now derive a model for portfolios of European options incorporating transaction costs. We can follow the Black–Scholes analysis up to equation (3.6), but in equation (3.7) we must allow for the cost of the transaction. If Π denotes the value of the hedged portfolio and $\delta\Pi$ the change in the portfolio over the time-step δt, then we must subtract the cost of any transaction from the right-hand side of the equation for $\delta\Pi$.

After a time-step the change in the value of the hedged portfolio is then

$$\begin{aligned} \delta\Pi = {} & \sigma S \left(\frac{\partial V}{\partial S} - \Delta \right) \phi \sqrt{\delta t} \\ & + \left(\tfrac{1}{2}\sigma^2 S^2 \frac{\partial^2 V}{\partial S^2} \phi^2 + \mu S \frac{\partial V}{\partial S} + \frac{\partial V}{\partial t} - \mu \Delta S \right) \delta t - \kappa S \mid \nu \mid . \end{aligned} \tag{16.1}$$

Here we have subtracted off the transaction costs (which are always positive, hence the modulus sign, $|\cdot|$, above). Since we have not gone to the limit $\delta t = 0$ we cannot replace the square of the random variable ϕ by its expected value, 1. Other than these differences the equation is the same as in Chapter 3.

Let us follow the same hedging strategy as before and choose $\Delta = \partial V/\partial S$. The number of assets held is therefore

$$\Delta = \frac{\partial V}{\partial S}(S,t)$$

where this has been evaluated at time t and asset value S. After a time-step δt and rehedging, the number of assets we hold becomes

$$\frac{\partial V}{\partial S}(S+\delta S, t+\delta t).$$

This is evaluated at the new time and asset price. On subtracting the former from the latter we find the number of assets we have traded to maintain a hedged position. This is

$$\nu = \frac{\partial V}{\partial S}(S+\delta S, t+\delta t) - \frac{\partial V}{\partial S}(S,t).$$

We can apply Taylor's theorem to expand the first term on the right-hand side for small δS and δt:

$$\frac{\partial V}{\partial S}(S+\delta S, t+\delta t) = \frac{\partial V}{\partial S}(S,t) + \delta S \frac{\partial^2 V}{\partial S^2}(S,t) + \delta t \frac{\partial^2 V}{\partial S \partial t}(S,t) + \cdots.$$

Since $\delta S = \sigma S \phi \sqrt{\delta t} + O(\delta t)$, the dominant term is that which is proportional to δS; this term is $O(\sqrt{\delta t})$ whereas the other terms are $O(\delta t)$. We find that to leading order the number of assets bought (sold) is

$$\nu \approx \frac{\partial^2 V}{\partial S^2}(S,t)\,\delta S \approx \frac{\partial^2 V}{\partial S^2}\sigma S \phi \sqrt{\delta t}.$$

Thus the expected transaction cost in a time-step is

$$\mathcal{E}\left[\kappa S |\nu|\right] = \sqrt{\frac{2}{\pi}}\kappa \sigma S^2 \left| \frac{\partial^2 V}{\partial S^2} \right| \sqrt{\delta t}. \tag{16.2}$$

(The factor $\sqrt{2/\pi}$ comes from calculating the expected value of $|\phi|$ using (2.3).) With our choice of Δ and with (16.2) as the expected transaction cost, we can calculate the expected change in the value of our portfolio from (16.1):

$$\mathcal{E}[\delta \Pi] = \left(\frac{\partial V}{\partial t} + \tfrac{1}{2}\sigma^2 S^2 \frac{\partial^2 V}{\partial S^2} - \kappa \sigma S^2 \sqrt{\frac{2}{\pi\,\delta t}} \left| \frac{\partial^2 V}{\partial S^2} \right| \right) \delta t. \tag{16.3}$$

Observe that, except for the modulus sign, the new term above, which is proportional to the transaction costs, is of the same form as the second S derivative that has appeared before.

If we assume that the holder of the option *expects* to make as much from his portfolio as if he had put the money in the bank, then we can replace the $\mathcal{E}[\delta\Pi]$ in (16.3) with $r(V - S\partial V/\partial S)\,\delta t$ as before to yield an equation for the value of the option:

$$\frac{\partial V}{\partial t} + \tfrac{1}{2}\sigma^2 S^2 \frac{\partial^2 V}{\partial S^2} - \kappa\sigma S^2 \sqrt{\frac{2}{\pi\,\delta t}} \left| \frac{\partial^2 V}{\partial S^2} \right| + rS\frac{\partial V}{\partial S} - rV = 0. \qquad (16.4)$$

The financial interpretation of the term that is not present in the usual Black–Scholes equation is clear if we recall the comments in the section on hedging. The second derivative of the option price with respect to the asset price is the gamma, $\Gamma = \partial^2 V/\partial S^2$. This is a measure of the degree of mishedging of the hedged portfolio, bearing in mind that the time-step is not infinitesimally small. The leading order component of randomness has been eliminated – this is delta-hedging – leaving behind a small component proportional to the gamma. Thus the gamma is related to the amount of rehedging that takes place in the next time interval and hence to the expected transaction costs.

The equation – which is a *nonlinear* parabolic partial differential equation, one of the few such in finance – is also valid for a portfolio of derivative products. This is the only time in this book that we notice any difference between single options and a portfolio of options. In the presence of transaction costs the value of a portfolio which is the sum of individual options is not the same as the sum of the values of the individual components. Thus it is to be expected that (16.4) is nonlinear. We can best see this by taking a very extreme case.

Suppose we have positions in two call options with the same exercise price and the same expiry date and on the same underlying asset. However, one of these is held long and the other short. Our net position is therefore zero. Our book of options is so large that we do not notice the cancellation effect of the two opposite positions and so decide to hedge each of them separately. Because of transaction costs we lose money at each rehedge on both options. At expiry we have a negative net balance, since the two payoffs cancel out but the costs remain. This contrasts greatly with our net balance at expiry if we realise that our positions are opposite. In the latter case we never rehedge, which leaves us with no transaction costs and a net balance of zero at expiry.

We give numerical results for a portfolio of options below. First, however, we consider the effect of costs on a single option held long. We know that

$$\frac{\partial^2 V}{\partial S^2} > 0$$

for a single call or put held long in the absence of transaction costs, as can be shown by differentiating (3.17) and (3.18). Let us postulate that this is true for a single call or put when transaction costs are present. We thus drop the modulus sign from (16.4) for the moment. With the notation

$$\breve{\sigma}^2 = \sigma^2 - 2\kappa\sigma\sqrt{\frac{2}{\pi\,\delta t}} \tag{16.5}$$

the equation for the value of the option is identical to the Black–Scholes value with the exception that the actual variance σ^2 is replaced by the modified variance $\breve{\sigma}^2$. Thus our assumption that $\partial^2 V/\partial S^2 > 0$ is true for a single vanilla option even in the presence of transaction costs. This is one way of valuing a long position on an option with transaction costs.

For a short option position we change all the signs in the above analysis with the exception of the transaction cost term, which must always be a drain on the portfolio. We then find that the option is valued using the new variance

$$\hat{\sigma}^2 = \sigma^2 + 2\kappa\sigma\sqrt{\frac{2}{\pi\,\delta t}}. \tag{16.6}$$

The results (16.5) and (16.6) show that a long position in a single call or put has an apparent volatility that is less than the actual volatility. This is because when the asset price rises the owner of the option must sell some assets to remain delta-hedged; however, the effect of the bid-offer spread on the underlying is to reduce the price at which the asset is sold and so the effective increase in the asset price is less than the actual increase. The converse is true for a short option position.

Staying with a single call or put, we can get some idea of the total transaction costs associated with the above strategy by examining the difference between the value of an option with the modified variance and that with the usual variance; that is, the difference between the value of the option taking into account the costs and the Black–Scholes value. Thus consider

$$V(S,t) - \hat{V}(S,t)$$

with the obvious notation. Expanding this expression for small κ we find that it becomes

$$\frac{\partial V}{\partial \sigma}(\sigma - \hat{\sigma}) + \cdots.$$

Since we know the formula for a European call option we find the expected spread to be

$$\frac{2\kappa S N'(d_1)\sqrt{(T-t)}}{\sqrt{2\pi\,\delta t}},$$

where $N(d_1)$ has its usual meaning.

Perhaps the most important quantity that appears in this model is

$$K = \frac{\kappa}{\sigma\sqrt{\delta t}}. \tag{16.7}$$

If $K \gg 1$ then the transaction costs term swamps the basic variance. This implies that costs are too high and that the chosen δt is too small. The portfolio is being rehedged too often.[1]

If $K \ll 1$ then the costs term affects only the basic variance marginally. This implies very low transaction costs. Hence δt is too large, and it should be decreased to minimise risk. The portfolio is being rehedged too seldom.

16.3 Portfolios of Options

We now consider the valuation of portfolios of options. For a general portfolio of options, the gamma, $\partial^2 V/\partial S^2$, is not of one sign. In this case we cannot drop the modulus sign. Since the problem is nonlinear we must in general solve equation (16.4) numerically, except in some very special cases, such as the vanilla option mentioned above, where the gamma does not change sign. The numerical solution is most easily achieved by an obvious generalisation of the explicit finite-difference method of Section 8.4 and exercise 9 of Chapter 8. The implicit methods of Chapter 8 can be used, but they require the solution of systems of nonlinear equations and this reduces their efficiency and relative advantage over the explicit method.

In Figures 16.1 and 16.2 we show the value of a long bullish vertical spread (one long call with $E = 45$ and one short call with $E = 55$) and the delta at six months before expiry for the two cases, with and without transaction costs. The volatility is 0.4 and the interest rate 0.1. The bold curve shows the values in the presence of transaction

[1] If the transaction costs are very large or the portfolio is rehedged very often, then it is possible to have $\kappa > 2\sigma\sqrt{2\delta t/\pi}$. If this is the case then the diffusion equation has a negative coefficient for a long option position and is thus ill-posed. This is because, although the asset price may have risen, its effective value due to the addition of the costs will have actually dropped.

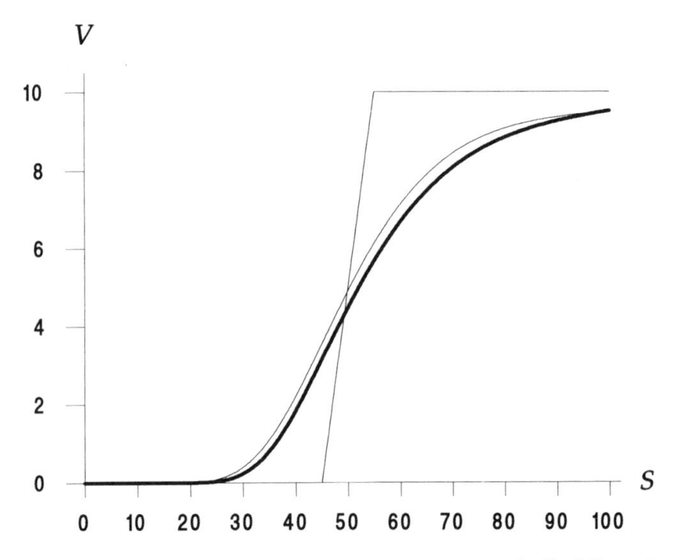

Figure 16.1 The value of a bullish vertical spread with (bold) and without transaction costs. The payoff is also shown.

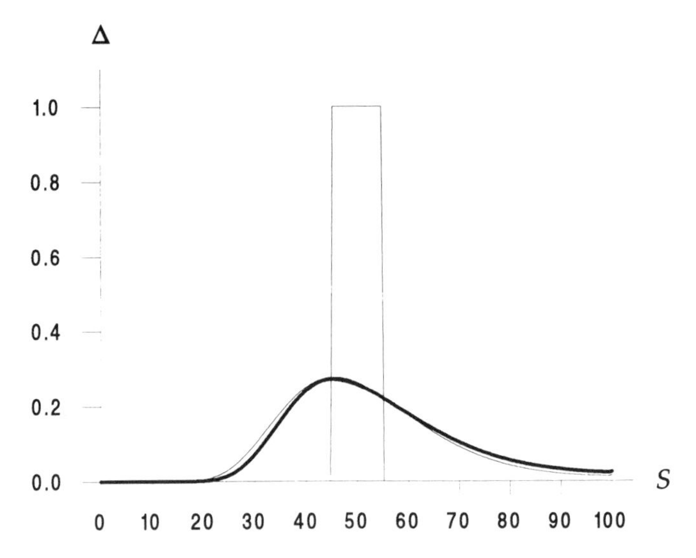

Figure 16.2 The delta for a bullish vertical spread prior to and at expiry with (bold) and without transaction costs.

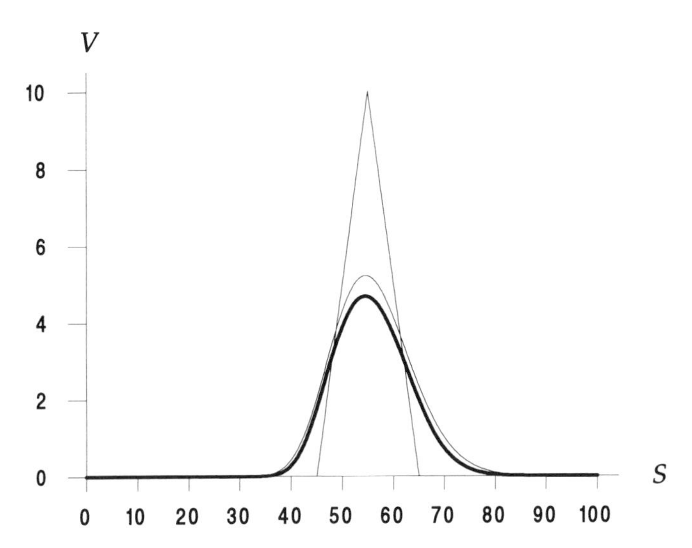

Figure 16.3 The value of a butterfly spread with (bold) and without transaction costs.

costs and the other curve in the absence of transaction costs. In this example $K = 0.25$. The latter is simply the Black–Scholes value for the combination of the two options. The bold line approaches the other line as the transaction costs decrease.

In Figures 16.3 and 16.4 we show the value of a long butterfly spread and its delta, before and at expiry. In this example the portfolio contains one long call with $E = 45$, two short calls with $E = 55$ and another long call with $E = 65$. The results are with one month until expiry for the two cases, with and without transaction costs. The volatility, the interest rate and K are as in the previous example.

Further Reading

- All the material of this chapter is based on the model of Leland (1985) as extended to portfolios of options by Hoggard, Whalley & Wilmott (1994).
- Boyle & Emanuel (1980) explain some of the problems associated with the *discrete* rehedging of an option portfolio.
- Gemmill (1992) gives an example taken from practice of the effect of transaction costs on a hedged portfolio.

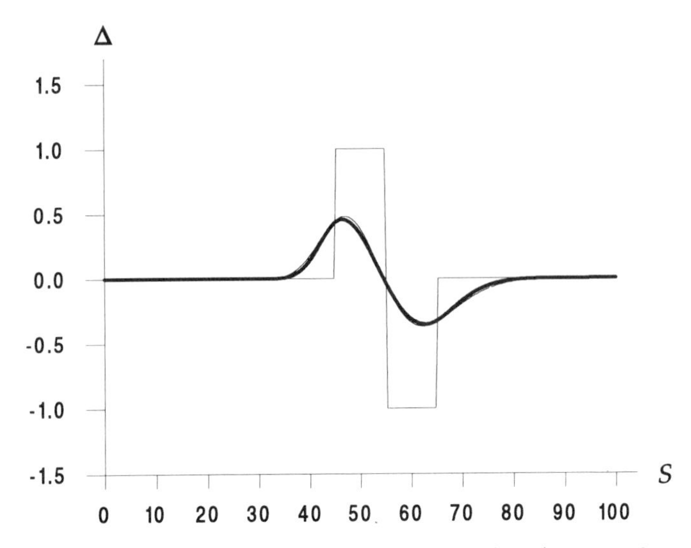

Figure 16.4 The delta for a butterfly spread with (bold) and without transaction costs.

- Whalley & Wilmott (1993) discuss various hedging strategies and derive more nonlinear equations using ideas similar to those in this chapter.
- For alternative approaches involving 'optimal strategies' see Hodges & Neuberger (1989), Davis & Norman (1990) and Davis, Panas & Zariphopoulou (1993), and the asymptotic analyses of Whalley & Wilmott (1993, 1994a,b) for small levels of transaction costs.
- Dewynne, Whalley & Wilmott (1994) discuss the pricing of exotic options in the presence of costs.

Exercises

1. Generalise the analysis of this chapter to include transaction costs which have three components: a fixed cost at each transaction, a cost proportional to the number of assets traded and a cost proportional to the value of the assets traded (only the last is included in our analysis here).

2. Continuing with the theme of a more general cost structure, show that under this general cost structure option prices can become negative. Is this financially reasonable? In the light of this, can the model be improved?

3. When the option's gamma $\partial^2 V/\partial S^2$ is of one sign it is simple to adjust the volatility to accommodate transaction costs. In other cases it is much harder to find explicit solutions. Consider instead, therefore, the simpler nonlinear diffusion equation

$$\frac{\partial u}{\partial t} = \frac{\partial^2 u}{\partial x^2} + \lambda \left| \frac{\partial^2 u}{\partial x^2} \right|.$$

Solve this equation for all x with the initial data

$$u(x, 0) = \delta(x).$$

4. If $K = \kappa/\sigma\sqrt{\delta t}$ is small then it is possible to solve (16.4) by iteration. What effect does this have on the nonlinear equation?

5. Suppose you hold a portfolio of options all expiring at time T. This portfolio has value $V(S, t)$ satisfying (16.4) with payoff $V(S, T)$ at expiry. The opportunity arises to issue a new contract having payoff $\Lambda(S)$ at time T. Find the equation satisfied by the *marginal* value of the new option, assuming that this new contract has only a small value compared with the initial portfolio.

Part four

Interest Rate Derivative Products

17 Interest Rate Models and Derivative Products

17.1 Introduction

One of the biggest assumptions we have so far made in this book is that interest rates are constant or, at least, known functions of time. In reality this is far from the case. Although the effects of interest rate changes on traded-option prices are relatively small, because of their short lifetime, many other securities that are also influenced by interest rates have much longer duration. Their analysis in the presence of unpredictable interest rates is of crucial practical importance. In the final part of this book, we give a brief introduction to pure interest rate derivative products, and then to products depending on both interest rates and an underlying asset.

We begin in this chapter with the subject of bond pricing. We do this first under the assumption of a deterministic interest rate. This simplification allows us to discuss the effect of coupons on the prices of bonds and the appearance of the yield curve, which we define shortly. Later in the chapter we relax the assumption of deterministic interest rates and present a model which allows the short-term interest rate, the spot rate, to follow a random walk. This leads to a parabolic partial differential equation for the prices of bonds and to models for bond options and many other interest rate derivative products.

17.2 Basics of Bond Pricing

A **bond** is a contract, paid for up-front, that yields a known amount on a known date in the future, the **maturity**[1] **date**, $t = T$. The bond may also pay a known cash dividend (the **coupon**) at fixed times during

[1] Convention has it that bonds 'mature' while options 'expire'.

the life of the contract. If there is no coupon the bond is known as a **zero-coupon bond**. Bonds may be issued by both governments or companies. The main purpose of a bond issue is the raising of capital, and the up-front premium can be thought of as a loan to the government or the company.

The problem of valuing a bond can be illustrated by the question

- How much should I pay now to get a guaranteed $1 in 10 years' time?

As with option models we aim to find the fair value of the contract. In this example the life-span of the bond is 10 years, in contrast to a typical equity option's life-span of nine months or less. For this reason the modelling of any time-dependent process must be more accurate when pricing bonds. It is not true, for instance, that interest rates remain constant for 10-year periods nor are they known in advance over such a period. However, we begin this chapter with the simplest model for a bond using the assumption that r is a known function of time. Having established a theoretical framework, we then introduce a model for stochastic interest rate movements.

17.2.1 Bond Pricing with Known Interest Rates

We continue to use the notation V to represent the value of the contract, in this case a bond. If the interest rate $r(t)$ and coupon payment $K(t)$ are known functions of time, the bond price is also a function of time only: $V = V(t)$. (The bond price is, of course, also a function of maturity date T, but we suppress that dependence except when it is important.) If this bond pays the owner Z at time $t = T$ then we know that $V(T) = Z$. We now derive an equation for the value of the bond at a time before maturity, $t < T$.

Suppose we hold one bond. The change in the value of that bond in a time-step dt (from t to $t + dt$) is

$$\frac{dV}{dt}\,dt.$$

If during this period we have received a coupon payment of $K(t)\,dt$, which may be either in the form of continuous or discrete payments or a combination, our holdings including cash change by an amount

$$\left(\frac{dV}{dt} + K(t)\right) dt.$$

Arbitrage considerations again lead us to equate this with the return from a bank deposit receiving interest at a rate $r(t)$. Thus we conclude that

$$\frac{dV}{dt} + K(t) = r(t)V; \tag{17.1}$$

the right-hand side is the return we would have received had we converted our bond into cash at time t. The solution of this ordinary differential equation is easily found with the help of the integrating factor $e^{-\int^t r(\tau)d\tau}$ to be

$$V(t) = e^{-\int_t^T r(\tau)d\tau}\left(Z + \int_t^T K(t')e^{\int_{t'}^T r(\tau)d\tau}dt'\right); \tag{17.2}$$

the arbitrary constant of integration has been chosen to ensure that $V(T) = Z$. Note that a positive coupon payment increases the value of the bond at time t.

Now suppose that there are zero-coupon bonds of all possible maturity dates, that is, there are bonds with $K(t) = 0$. Still supposing that the interest rate is deterministic, we have

$$V(t;T) = Ze^{-\int_t^T r(\tau)d\tau}, \tag{17.3}$$

from (17.2) with $K = 0$ (we have now made the dependence on T explicit). If the bond prices are quoted today, at time t, for all values of the maturity date T then we know the left-hand side of (17.3) for all values of T. Thus

$$-\int_t^T r(\tau)\,d\tau = \log\left(V(t;T)/Z\right). \tag{17.4}$$

If $V(t;T)$ is differentiable with respect to T, then differentiating (17.4) we have

$$r(T) = \frac{-1}{V(t;T)}\frac{\partial V}{\partial T}. \tag{17.5}$$

If the market prices of the zero-coupon bonds genuinely reflect a known, deterministic interest rate then that interest rate at future dates is given from the bond prices by (17.5). Since the interest rate r is positive we must have

$$\frac{\partial V}{\partial T} < 0.$$

Thus the longer a bond has to live, the less it is now worth; this result is financially clear.

17.2.2 Discretely Paid Coupons

Equation (17.8) allows for the payment of a coupon. But what if the coupon is paid discretely, as it is in practice, for example, every six months? We can arrive at this result by a financial argument that will be useful later. Since the holder of the bond receives K_c at time t_c, there must be a jump in the value of the bond across the coupon date (if not, there is an obvious arbitrage opportunity). That is, the value before and after this date differs by K_c:

$$V(t_c^-) = V(t_c^+) + K_c.$$

This will be recognised as a jump condition. This time the realised bond price is *not* continuous. After all, there is a discrete payment at the coupon date. This jump condition will still apply when we come to consider stochastic interest rates.

A more mathematical approach is to write $K(t)$ as the sum of delta functions, one for each coupon payment. Suppose, for simplicity, that there is just one payment of K_c at time $t_c < T$. We then have

$$\frac{dV}{dt} + K_c\delta(t - t_c) = r(t)V. \tag{17.6}$$

We can substitute the delta function into (17.2) to get

$$V(t) = e^{-\int_t^T r(\tau)d\tau}\left(Z + K_c\mathcal{H}(t_c - t)e^{\int_{t_c}^T r(\tau)d\tau}\right).$$

17.3 The Yield Curve

Despite all our assumptions to the contrary, interest rates are not deterministic. For short-dated derivative products such as options the errors associated with assuming a deterministic, or even constant, rate are small, typically 2%. In dealing with products with a longer lifespan we must address the problem of random interest rates. The first step is to decide on a suitable measure for future values of interest rates, one that enables traders to communicate effectively about the same quantity. In the previous section we have seen a definition, (17.5), that gives an interest rate from bond price data but this relies on bond prices being differentiable with respect to the maturity date.

The **yield curve** is another measure of future values of interest rates. With the value of zero-coupon bonds $V(t;T)$ taken from real data, define

$$Y(t;T) = -\frac{\log\left(V(t;T)/V(T;T)\right)}{T-t}, \tag{17.7}$$

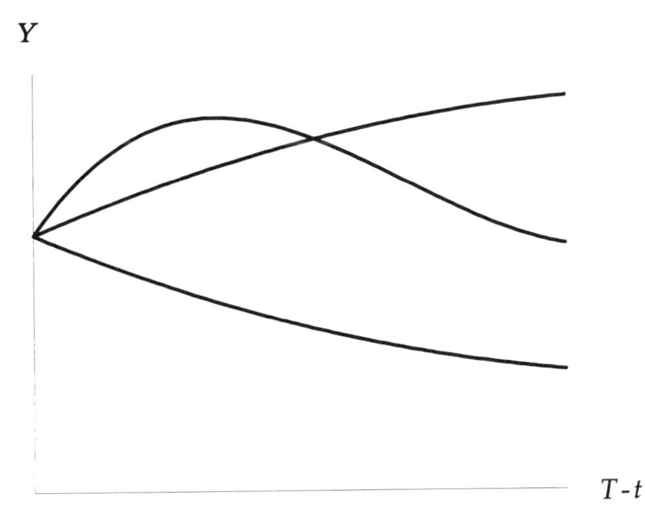

Figure 17.1. Typical yield curves: increasing, decreasing and humped.

where t is the current time. The yield curve is the plot of Y against time to maturity $T-t$. The dependence of the yield curve on the time to maturity is called the **term structure of interest rates**. This definition for Y has advantages over the measure (17.5), since

- bond prices, $V(t;T)$, do not have to be differentiable;
- a continuous distribution of bonds with all maturities is not required.

Y has the same dimensions as interest rates, i.e. inverse time. The two measures of future interest rates, (17.5) and (17.7), are the same when interest rates are constant.

It is observed from market data that yield curves typically come in three distinct shapes, each associated with different economic conditions:

- increasing – this is the most common form for the yield curve. Future interest rates are higher than the short-term rate, since it should be more rewarding to tie money up for a long time than for a short time;
- decreasing – this is typical of periods when the short rate is high but expected to fall;
- humped – again the short rate is expected to fall.

These are all illustrated in Figure 17.1.

17.4 Stochastic Interest Rates

In view of our uncertainty about the future course of the interest rate, it is natural to model it as a random variable. For the rest of this chapter we assume that this is the case. To be technically correct we should specify that r is the interest rate received by the shortest possible deposit. If one is willing to tie money up for a long period of time then usually one gets a higher overall rate to offset the risk that the short rate will rise rapidly. (This would give an upward sloping yield curve.) The interest rate for the shortest possible deposit is commonly called the **spot rate**.

The subject of modelling interest rates is still in its infancy and we do not have the space here to discuss it in any depth. For these reasons we simply quote a fairly general interest rate model which, for reasons we mention below, has become popular.

In the same way that we proposed a model for the asset price as a lognormal random walk let us suppose that the interest rate r is governed by a stochastic differential equation of the form

$$dr = w(r,t)\,dX + u(r,t)\,dt. \tag{17.8}$$

The functional forms of $w(r,t)$ and $u(r,t)$ determine the behaviour of the spot rate r. We use this random walk to derive a partial differential equation for the price of a bond in a way similar to our derivation of the Black–Scholes equation. Later we choose functional forms for u and w that give a reasonable model for the spot rate.

17.5 The Bond Pricing Equation

When interest rates follow the stochastic differential equation (17.8) a bond has a price of the form $V(r,t)$; the dependence on T will be made explicit only when necessary.

Pricing a bond is technically harder than pricing an option, since *there is no underlying asset with which to hedge*: one cannot go out and 'buy' an interest rate of 5%. In this situation the only alternative is to hedge with bonds of different maturity dates. For this reason we set up a portfolio containing *two* bonds with different maturities, T_1 and T_2. The bond with maturity T_1 has price V_1 and the bond with maturity T_2 has price V_2. We hold one of the former and a number $-\Delta$ of the latter. Thus

$$\Pi = V_1 - \Delta V_2. \tag{17.9}$$

The change in this portfolio in a time dt is

$$d\Pi = \frac{\partial V_1}{\partial t}dt + \frac{\partial V_1}{\partial r}dr + \tfrac{1}{2}w^2\frac{\partial^2 V_1}{\partial r^2}dt - \Delta\left(\frac{\partial V_2}{\partial t}dt + \frac{\partial V_2}{\partial r}dr + \tfrac{1}{2}w^2\frac{\partial^2 V_2}{\partial r^2}dt\right), \tag{17.10}$$

where we have applied Itô's lemma to functions of r and t.

From (17.10) we see that the choice

$$\Delta = \frac{\partial V_1}{\partial r}\bigg/\frac{\partial V_2}{\partial r}$$

eliminates the random component in $d\Pi$. We then have

$$\begin{aligned} d\Pi &= \left(\frac{\partial V_1}{\partial t} + \tfrac{1}{2}w^2\frac{\partial^2 V_1}{\partial r^2} - \frac{\partial V_1/\partial r}{\partial V_2/\partial r}\left(\frac{\partial V_2}{\partial t} + \tfrac{1}{2}w^2\frac{\partial^2 V_2}{\partial r^2}\right)\right)dt \\ &= r\left(V_1 - \frac{\partial V_1}{\partial r}\bigg/\frac{\partial V_2}{\partial r}\,V_2\right)dt, \\ &= r\Pi\,dt \end{aligned}$$

where we have used arbitrage arguments to set the return on the portfolio equal to the risk-free rate, the spot rate.

Gathering together all V_1 terms on the left-hand side and all V_2 terms on the right-hand side we find that

$$\left(\frac{\partial V_1}{\partial t} + \tfrac{1}{2}w^2\frac{\partial^2 V_1}{\partial r^2} - rV_1\right)\bigg/\frac{\partial V_1}{\partial r} = \left(\frac{\partial V_2}{\partial t} + \tfrac{1}{2}w^2\frac{\partial^2 V_2}{\partial r^2} - rV_2\right)\bigg/\frac{\partial V_2}{\partial r}.$$

This is one equation in two unknowns. However, the left-hand side is a function of T_1 but not T_2, and the right-hand side is a function of T_2 but not T_1. The only way for this to be possible is for both sides to be independent of the maturity date. Thus, dropping the subscript from V,

$$\left(\frac{\partial V}{\partial t} + \tfrac{1}{2}w^2\frac{\partial^2 V}{\partial r^2} - rV\right)\bigg/\frac{\partial V}{\partial r} = a(r,t)$$

for some function $a(r,t)$. In view of later developments it is convenient to write

$$a(r,t) = w(r,t)\lambda(r,t) - u(r,t);$$

for given $w(r,t)$ (not identically zero) and $u(r,t)$ this is always possible. The function $\lambda(r,t)$ is as yet unspecified.

The zero-coupon bond pricing equation is therefore

$$\frac{\partial V}{\partial t} + \tfrac{1}{2}w^2\frac{\partial^2 V}{\partial r^2} + (u - \lambda w)\frac{\partial V}{\partial r} - rV = 0. \tag{17.11}$$

In order to solve (17.11) uniquely we must pose one final and two boundary conditions. The final condition corresponds to the payoff on maturity and so

$$V(r,T) = Z.$$

Boundary conditions depend on the form of $u(r,t)$ and $w(r,t)$ and are discussed later for a special model.

It is a simple matter to incorporate coupon payments into the model. The result is

$$\frac{\partial V}{\partial t} + \tfrac{1}{2}w^2\frac{\partial^2 V}{\partial r^2} + (u - \lambda w)\frac{\partial V}{\partial r} - rV + K = 0,$$

where K is the coupon payment and may be a function of r and t. We leave the demonstration of this as an exercise for the reader.

When this coupon is paid discretely we can write $K(t)$ as a sum of delta functions. From the mathematical and financial arguments of Section 17.2.2 we find that if there is a discrete payment of K_c at time t_c then $V(r,t)$ must satisfy the jump condition

$$V(r,t_c^-) = V(r,t_c^+) + K_c.$$

17.5.1 The Market Price of Risk

We can now give an elegant interpretation of the hitherto mysterious function $\lambda(r,t)$. Instead of holding the hedged portfolio that we constructed above, suppose that we hold just one bond with maturity date T. In a time-step dt this bond changes in value by

$$dV = w\frac{\partial V}{\partial r}dX + \left(\frac{\partial V}{\partial t} + \tfrac{1}{2}w^2\frac{\partial^2 V}{\partial r^2} + u\frac{\partial V}{\partial r}\right)dt.$$

From (17.11) this may be written as

$$dV = w\frac{\partial V}{\partial r}dX + \left(w\lambda\frac{\partial V}{\partial r} + rV\right)dt,$$

or

$$dV - rV\,dt = w\frac{\partial V}{\partial r}(dX + \lambda\,dt). \qquad (17.12)$$

The presence of dX in (17.12) shows that this is not a riskless portfolio. The right-hand side may be interpreted as the excess return above the risk-free rate for accepting a certain level of risk. In return for taking the extra risk the portfolio profits by an extra $\lambda\,dt$ per unit of extra risk,

dX. For this reason the function λ is often called the **market price of risk.**

Technical Point: The Market Price of Risk for Assets.
In Chapter 3 we derived the Black–Scholes equation by constructing a portfolio with one option and a number $-\Delta$ of the underlying asset. Suppose that instead we were to follow the analysis above with a portfolio of two options with different maturity dates (or different exercise prices, for that matter), so that

$$\Pi = V_1 - \Delta V_2.$$

As above, we find that

$$\frac{\partial V}{\partial t} + \tfrac{1}{2}\sigma^2 S^2 \frac{\partial^2 V}{\partial S^2} + (\mu - \lambda_S \sigma) S \frac{\partial V}{\partial S} - rV = 0; \qquad (17.13)$$

this is simply (17.11) with S instead of r, μS instead of u, λ_S instead of λ and σS instead of w. Now recall that hedging options is easier than hedging bonds because of the existence of a tradable underlying asset. In other words, $V = S$ must itself be a solution of (17.13). Substituting $V = S$ into (17.13) we find that

$$(\mu - \lambda_S \sigma) S - rS = 0,$$

i.e.

$$\lambda_S = \frac{\mu - r}{\sigma};$$

this is the market price of risk for assets. Now putting $\lambda_S = (\mu - r)/\sigma$ into (17.13) we arrive at

$$\frac{\partial V}{\partial t} + \tfrac{1}{2}\sigma^2 S^2 \frac{\partial^2 V}{\partial S^2} + rS \frac{\partial V}{\partial S} - rV = 0;$$

this is the Black–Scholes equation, which contains no reference to μ or λ_S.

17.6 Solutions of the Bond Pricing Equation

Experience shows that the coefficients in (17.8) must have a more complicated form than the rather simple coefficients in the basic random walk for asset prices that we have used so far. Compensating for this is the fact that in the yield curve we have some detailed information about the behaviour of r, and it is important to be able to use this data effectively. The solution of the 'inverse problem', namely to derive the random walk from the yield curve, is much easier if we have explicit formulæ that relate bond prices to interest rates. Thus, we consider only certain special functional forms for w and u, which can be shown to be the most general such forms compatible with a particularly tractable class of solutions of the bond pricing equation.

We assume that w and u have the form

$$w(r,t) = \sqrt{\alpha(t)r - \beta(t)}, \tag{17.14}$$

$$u(r,t) = \left(-\gamma(t)r + \eta(t) + \lambda(r,t)\sqrt{\alpha(t)r - \beta(t)}\right). \tag{17.15}$$

The functions α, β, γ, η and λ that appear in (17.14) and (17.15) are functions of time. They are at our disposal to fit the data as well as possible; note, though, that the special form of (17.14) and (17.15) means that $\lambda(r,t)$ does not appear explicitly in the bond-pricing equation (17.11). By suitably restricting these time-dependent functions, we can ensure that the random walk (17.8) for r has the following economically plausible properties:

- We can avoid negative interest rates: the spot rate can be bounded below by a positive number if we insist that $\alpha(t) > 0$ and $\beta \geq 0$. The lower bound is then β/α. (In the special case $\alpha(t) = 0$ we must take $\beta(t) \leq 0$.) Note that r can still go to infinity, albeit with probability zero.
- We can make the spot rate **mean reverting**. For large (small) r the interest rate will tend to decrease (increase) towards the mean, which may be a function of time.

In addition, we require that if r reaches its lower bound β/α, it thereafter increases. This requirement can be shown to force

- $\eta(t) \geq \beta(t)\gamma(t)/\alpha(t) + \alpha(t)/2$,

and it is discussed further below.

There are many interest rate models, associated with the names of their inventors. The stochastic differential equation (17.8) incorporates the models of

- Vasicek ($\alpha = 0$, no time dependence in the parameters);
- Cox, Ingersoll & Ross ($\beta = 0$, no time dependence in the parameters);
- Hull & White (either $\alpha = 0$ or $\beta = 0$ but all parameters time-dependent).

With the model (17.14) and (17.15) we can state that the boundary conditions for (17.11) are, first, that

$$V(r,t) \to 0 \quad \text{as} \quad r \to \infty,$$

and, second, that on $r = \beta/\alpha$, V remains finite.[2]

Why did we choose u and w to take the forms (17.14) and (17.15)? All of the models mentioned above take special functional forms for the coefficients of dt and dX in the stochastic differential equation for r, so that the solution of (17.11) is of the simple form[3]

$$V(r,t) = Ze^{A(t;T)-rB(t;T)}. \tag{17.16}$$

(We have explicitly shown the dependence of V, A and B on T.) Indeed the model with all of α, β, γ and η non-zero is the most general stochastic differential equation for r which leads to a solution of (17.11) of the form (17.16).

That this is so is easily shown. First substitute (17.16) into the bond pricing equation (17.11). This gives

$$\frac{\partial A}{\partial t} - r\frac{\partial B}{\partial t} + \tfrac{1}{2}w^2B^2 - (u - \lambda w)B - r = 0. \tag{17.17}$$

Some of these terms are functions of t and T (i.e. A and B) and others are functions of r and t (i.e. u and w). Differentiating (17.17) with respect to r gives

$$-\frac{\partial B}{\partial t} + \tfrac{1}{2}B^2\frac{\partial(w^2)}{\partial r} - B\frac{\partial(u-\lambda w)}{\partial r} = 0.$$

After a further differentiation with respect to r, and after dividing through by B, we get

$$\tfrac{1}{2}B\frac{\partial^2(w^2)}{\partial r^2} - \frac{\partial^2(u-\lambda w)}{\partial r^2} = 0.$$

Since B is a function of T we must have

$$\frac{\partial^2(w^2)}{\partial r^2} = 0 \tag{17.18}$$

[2] When r is bounded below by β/α, a local analysis of the partial differential equation can be carried out near $r = \beta/\alpha$ (see the exercises at the end of this chapter). Briefly, balancing the terms $\frac{1}{2}(\alpha r - \beta)\partial^2 V/\partial r^2$ and $(\eta - \gamma r)\partial V/\partial r$ shows that finiteness of V at $r = \beta/\alpha$ is a sufficient boundary condition only if $\eta \geq \beta\gamma/\alpha - \alpha/2$.

[3] The existence of a simple explicit solution, while being an obvious advantage in a model, is not usually a good reason for accepting a model as representative of the real world. However, the *inverse* nature of fitting yield curves makes these models useful.

and

$$\frac{\partial^2(u - \lambda w)}{\partial r^2} = 0. \tag{17.19}$$

Thus we arrive at (17.14) and (17.15) as the choices for u and w.

We omit the details, but the substitution of (17.14) and (17.15) into (17.18) and (17.19) and equating powers of r yields the following equations for A and B:

$$\frac{\partial A}{\partial t} = \eta(t)B + \tfrac{1}{2}\beta(t)B^2 \tag{17.20}$$

and

$$\frac{\partial B}{\partial t} = \tfrac{1}{2}\alpha(t)B^2 + \gamma(t)B - 1. \tag{17.21}$$

In order to satisfy the final data that $V(r,T) = Z$ we must have

$$A(t;T) = 0 \quad \text{and} \quad B(t;T) = 0.$$

17.6.1 Analysis for Constant Parameters

The solution for arbitrary α, β, γ and η is found by integrating the two ordinary differential equations (17.20) and (17.21). Generally this cannot be done explicitly, but a simple case is when α, β, γ and η are all constant. In this case it is found that

$$\frac{2}{\alpha}A = a\psi_2 \log(a - B) + (\psi_2 - \tfrac{1}{2}\beta)b\log((B+b)/b) + \tfrac{1}{2}B\beta - a\psi_2 \log a, \tag{17.22}$$

and

$$B(t;T) = \frac{2(e^{\psi_1(T-t)} - 1)}{(\gamma + \psi_1)(e^{\psi_1(T-t)} - 1) + 2\psi_1}, \tag{17.23}$$

where

$$b, a = \frac{\pm\gamma + \sqrt{\gamma^2 + 2\alpha}}{\alpha},$$

and

$$\psi_1 = \sqrt{\gamma^2 + 2\alpha} \quad \text{and} \quad \psi_2 = \frac{\eta + a\beta/2}{a+b}.$$

When all four of the parameters are constant it is obvious that both A and B are functions of only the one variable $\tau = T - t$; this would not necessarily be the case if any of the parameters were time-dependent.

A wide variety of yield curves can be predicted by the model, including increasing, decreasing and humped. As $\tau \to \infty$,

$$B \to \frac{2}{\gamma + \psi_1}$$

and the yield curve Y has long-term behaviour given by

$$Y \to \frac{2}{(\gamma + \psi_1)^2} \left(\eta(\gamma + \psi_1) + \beta \right).$$

Thus for constant and fixed parameters the model leads to a fixed long-term interest rate, independent of the spot rate.

17.6.2 Fitting the Parameters

The general stochastic process developed in this chapter involves four time-dependent parameters, α, β, γ and η. If these parameters are assumed to be constant then explicit forms for A, B and hence bond prices are easily obtained.

However, it is reasonable to conjecture that the market's expectations about future interest rates are time-varying. This time dependence may, for example, arise from the cyclical nature of the economy. We now give an overview of a possible approach to incorporating one time-dependent parameter in the general model while the other three parameters are kept constant. When introducing time-dependent parameters, careful consideration must be given to what information is available from, and relevant to, the market.

The methodology of this section is as follows. We insist that α, β and γ are constant and allow η to be a function of time. We see that this gives sufficient freedom to fit the market yield curve exactly.

The first step is to determine α and β. There is sufficient information in historic data to find these if we know the lower bound for interest rates and the volatility of the spot rate. Having determined α and β we then go on to find γ. This is found by considering the correlation between the spot rate and the slope of the yield curve. Finally, the function $\eta(t)$ is chosen to fit the full yield curve exactly. This involves the solution of an integral equation.

Bounding r Below

Suppose that we are interested in valuing a 10-year bond. It is possible that an investor has a view about the likely lower bound for interest rates over the next 10 years. Alternatively, it may be reasonable to postulate that a lower bound for interest rates over this period is similar to the smallest value achieved in the past 10 years. This is analogous to using the historic volatility over a period comparable to the life of the option as a volatility measure in option pricing. In any event, let us suppose that a lower bound has been decided on. In this case the quantity β/α is 'known'.

The Spot Rate Volatility

The spot rate volatility is simply

$$\sqrt{\alpha r - \beta}.$$

Again, this is easy to estimate, if it is assumed not to be time-dependent. Thus from the historic lower bound and the current volatility we have sufficient information to estimate both α and β.

The Volatility of the Yield Curve Slope

It is easy to solve (17.20) and (17.21) by Taylor series for values of t close to T. Such an analysis shows that the yield curve Y, which is now given by

$$Y = \frac{-A + rB}{T - t},$$

can be approximated for times close to maturity by

$$Y \sim r - \tfrac{1}{2}(T-t)\Big(\gamma r - \eta(0)\Big) + \cdots.$$

We can see from this that the slope of the yield curve at the short end (i.e. at $T = t$) is given by

$$s = \tfrac{1}{2}\Big(\eta(0) - \gamma r\Big). \qquad (17.24)$$

Note that this model predicts that this slope depends on the spot rate itself.

If the spot rate is indeed mean-reverting, which is the case if $\gamma(t)$ is positive, an increase in the spot rate r leads to a decrease in the slope (17.24): if the spot rate increases, the yield curve flattens. Moreover, as the spot rate follows a random walk, so does the slope of the short rate. Since the two are linked by the deterministic equation (17.24), these changes are perfectly correlated:

$$ds = -\tfrac{1}{2}\gamma\, dr.$$

An examination of the data for r and s gives γ as minus twice the covariance of ds and dr divided by the variance of dr. It may happen that the data give a negative value for γ, so that the spot rate random walk and the local yield curve slope random walk are positively correlated: if the spot rate drops then the yield curve steepens. This is indicative of a spot rate which is not mean-reverting.

The Whole Yield Curve

So far we have fitted the constant parameters α, β and γ in a simple and practical manner. We now come to choose $\eta(t)$ so as to fit the term structure in the market exactly. We see that this leads to an integral equation for $\eta(t)$ which must be solved numerically except in simple cases.

We can integrate (17.20) explicitly to find that

$$A = -\tfrac{1}{2}\beta \int_t^T B^2(T-s)\, ds - \int_t^T \eta(s) B(T-s)\, ds \qquad (17.25)$$

where $B(T-t)$ is given by (17.23) with the obvious notation; it is only a function of $T-t$. This expression is known exactly except for the final integral term involving $\eta(t)$.

Suppose that we wish to fit the yield curve once only, at time t^*. At this time the spot rate is r^*, the yield curve, which is known from market data, is $Y^*(T)$ and the four parameters in the model are denoted by α^*, β^*, γ^* and η^*. We are using asterisks to denote the values of the parameters and data at the point in time t^*.

If we now substitute the known yield curve into (17.25) we find that $\eta^*(t)$ satisfies the integral equation

$$\int_{t^*}^T \eta^*(s) B(T-s)\, ds = Br^* - Y^*(T-t^*) - \tfrac{1}{2}\beta^* \int_{t^*}^T B^2(T-s)\, ds. \qquad (17.26)$$

This must be solved for $t^* \leq T < \infty$. Once the solution of this equation has been found we know all of α^*, β^*, γ^* and $\eta^*(t)$. Substitution of these into (17.23) gives the expression for B and then into (17.25) gives A. The price of any bond is then given by

$$Ze^{A(t;T)-rB(t;T)}.$$

It is possible to solve (17.26) exactly by taking Laplace transforms, since the equation is of Volterra type with a convolution kernel. Unfortunately, B does not have a simple transform and thus this method is impractical. Fortunately, the integral equation is not difficult to solve numerically, but we do not discuss this any further here.

We have fitted the yield curve exactly at time t^*. In so doing we have found the three constant parameters α, β and γ and the time-dependent parameter η. The model is only strictly valid if, when we come to refit these parameters at a later date, they remain the same. This is unlikely to be the case and is because the basic model (17.8) has been chosen for its analytic properties and not on the basis of any economic modelling. This is a weakness of many currently popular models.

17.7 The Extended Vasicek Model of Hull & White

In the Vasicek (1977) model as extended to include time-dependent parameters by Hull & White (1990), $\alpha(t) = 0$ and $\beta < 0$. Although Hull & White advocate a very sophisticated choice of $\beta(t)$, $\gamma(t)$ and $\eta(t)$ (all time-dependent) to fit spot rate volatility, yield curve volatility for all maturities etc., we allow only η to be time-dependent as suggested above.

In this model $\alpha = 0$, and we assume that β^* and γ^* have been determined at time t^*. In this case $B(T-t)$ simplifies to

$$B(T-t) = \frac{1}{\gamma^*}\left(1 - e^{-\gamma^*(T-t)}\right)$$

and the integral equation for η^* becomes

$$\int_{t^*}^{T} \eta^*(s)\left(1 - e^{-\gamma^*(T-s)}\right) ds = \gamma^* F^*(T). \qquad (17.27)$$

Here F^* is a known function of T, given by the right-hand side of (17.26), which in particular depends on integrals of B and the current state of the yield curve. For this integral equation to have a solution we must have $F(0) = 0$. That this is indeed the case can be seen from the right-hand side of (17.26).

Although (17.27) may be solved by Laplace transform methods as suggested above, it is particularly easy to solve by differentiating the equation twice with respect to T. After the first differentiation we have

$$\int_{t^*}^{T} \eta^*(s) e^{-\gamma^*(T-s)} ds = F^{*\prime}(T),$$

where $'$ denotes d/dT. The second differentiation gives

$$\eta^*(T) - \gamma^* \int_{t^*}^{T} \eta^*(s) e^{-\gamma^*(T-s)} ds = F^{*\prime\prime}(T).$$

We can eliminate the integrals between these two expressions to find that

$$\eta^*(T) = F^{*\prime\prime}(T) + \gamma^* F^{*\prime}(T).$$

The expression for $\eta^*(T)$ is

$$\eta^*(T) = -Y^{*\prime\prime}(T) - \gamma^* Y^{*\prime}(T) - \beta^*(T - t^*) - \frac{\beta^*}{2\gamma^*}\left(1 - e^{-2\gamma^*(T-t^*)}\right). \tag{17.28}$$

From this expression we can now find the function $A(t;T)$ using (17.20). We leave this simple task as an exercise for the reader.

17.8 Bond Options

The theoretical model for the spot rate presented above allows us to value contingent claims such as bond options. A **bond option** is identical to an equity option except that the underlying asset is a bond. Both European and American versions exist.

As a simple example, we derive the differential equation satisfied by a call option, with exercise price E and expiry date T, on a zero-coupon bond with maturity date $T_B \geq T$. Before finding the value of the option to buy a bond we must find the value of the bond itself.

Let us write $V_B(r, t; T_B)$ for the value of the bond. Thus, V_B satisfies

$$\frac{\partial V_B}{\partial t} + \tfrac{1}{2} w^2 \frac{\partial^2 V_B}{\partial r^2} + (u - \lambda w) \frac{\partial V_B}{\partial r} - r V_B = 0 \tag{17.29}$$

with

$$V_B(r, T_B, T_B) = Z$$

and suitable boundary conditions. Now write $C_B(r, t)$ for the value of the call option on this bond. Since C_B also depends on the random variable r, it too must satisfy equation (17.29). The only difference is that the final value for the option is

$$C_B(r, T) = \max(V_B(r, t; T_B) - E, 0).$$

This idea can readily be generalised, as we now see.

17.9 Other Interest Rate Products

There is a large, and ever-growing, number of different interest rate derivative products. We do not have the space to include any but the most common. However, the following examples show the way in which many such products can be incorporated into the same partial differential equation framework.

17.9.1 Swaps

An interest rate **swap** is an agreement between two parties to exchange the interest rate payments on a certain amount, the principal, for a certain length of time. One party pays the other a fixed rate of interest in return for a variable interest rate payment. For example, A pays 9% of $1,000,000 p.a. to B and B pays r of the same amount to A. This agreement is to last for three years, say. We now value such swaps in general.

Suppose that A pays the interest on an amount Z to B at a fixed rate r^* and B pays interest to A at the floating rate r. These payments continue until time T. Let us denote the value of this swap to A by $ZV(r,t)$.

We can accommodate this product into our interest rate framework by observing that in a time-step dt A receives $(r-r^*)Z\,dt$. If we think of this payment as being similar to a coupon payment on a simple bond then we find that

$$\frac{\partial V}{\partial t}+\tfrac{1}{2}w^2\frac{\partial^2 V}{\partial r^2}+(u-\lambda w)\frac{\partial V}{\partial r}-rV+(r-r^*)=0,$$

with final data

$$V(r,T)=0.$$

Note that r can be greater or less than r^* and so $V(r,t)$ need not be positive. Indeed, a swap is not necessarily an asset but can be a liability depending on, for example, the current state of the yield curve.

17.9.2 Caps and Floors

A **cap** is a loan at the floating interest rate but with the proviso that the interest rate charged is guaranteed not to exceed a specified value, the **cap**, which we denote by r^*. The loan of Z is to be paid back at time

T. We readily find that the value of the capped loan, $ZV(r,t)$, satisfies

$$\frac{\partial V}{\partial t} + \tfrac{1}{2}w^2\frac{\partial^2 V}{\partial r^2} + (u - \lambda w)\frac{\partial V}{\partial r} - rV + \min(r, r^*) = 0, \qquad (17.30)$$

with

$$V(r,T) = 1.$$

A **floor** is similar to a cap except that the interest rate does not go *below* r^*. To value this contract simply replace $\min(r, r^*)$ by $\max(r, r^*)$ in (17.30).

17.9.3 Swaptions, Captions and Floortions

Having valued swaps, caps and floors it is easy to value options on these instruments: **swaptions**, **captions** and **floortions**. Suppose that our swap (cap or floor) which expires at time T_S has value $V_S(r,t)$ for $t \le T_S$. An option to buy this swap (a call swaption) for an amount E at time $T < T_S$ has value $V(r,t)$ where

$$\frac{\partial V}{\partial t} + \tfrac{1}{2}w^2\frac{\partial^2 V}{\partial r^2} + (u - \lambda w)\frac{\partial V}{\partial r} - rV = 0,$$

with

$$V(r,T) = \max\left(V_S(r,T) - E, 0\right).$$

Thus we solve for the value of the swap itself first, and then use this value as the final data for the value of the swaption. Captions and floortions are treated similarly.

Further Reading

We have given only a brief account of the rapidly developing subject of interest rate modelling. The interested reader will find many more details in the original papers listed below.

- See the original papers by Dothan (1978), Vasicek (1977), Cox, Ingersoll & Ross (1985), Ho & Lee (1986) and Black, Derman & Toy (1990). For details of the general model see the papers by by Pearson & Sun (1989), Duffie (1992), Klugman (1992) and Klugman & Wilmott (1993).
- A more sophisticated choice of time-dependent parameters is described by Hull & White (1990).

- Klugman & Wilmott (1993) solve the integral equation asymptotically for small α. Baker (1977) gives details of numerical treatments for integral equations.
- See Hull (1993) for details of other interest rate products and their uses in practice.

Exercises

1. From expression (17.28) find the function $A(t;T)$ using (17.20).
2. Verify the local analysis of the bond pricing equation near $r = \beta/\alpha$.
3. Suppose that a bond pays a coupon $K(r,t)$. Show that the bond pricing equation is modified to

$$\frac{\partial V}{\partial t} + \tfrac{1}{2}w^2\frac{\partial^2 V}{\partial r^2} + (u - \lambda w)\frac{\partial V}{\partial r} - rV + K = 0.$$

4. In practice a swap contract entails the exchange of interest payments at discrete times, usually every quarter. How does this affect the swap pricing partial differential equation?
5. Consider a swap having one discrete exchange of payments at time T only. This contract satisfies

$$\frac{\partial V}{\partial t} + \tfrac{1}{2}w^2\frac{\partial^2 V}{\partial r^2} + (u - \lambda w)\frac{\partial V}{\partial r} - rV = 0$$

with

$$V(r,T) = r - r^*.$$

The final condition represents an exchange of fixed and floating payments. Find explicit solutions to this problem for each of the interest rate models we have described here.

6. Find the Taylor series expansion of the zero-coupon bond about the maturity date for the case

$$dr = w(r)dX + u(r)dt.$$

7. Perform a Taylor series expansion of the swap with a single exchange around the time of this exchange.
8. It is common market practice to use the Black–Scholes formulæ to value options on bonds. What are the advantages and disadvantages of such an approach?

9. Suppose that an interest rate derivative has the payoff at $t = T$

$$\max(r - r^*, 0).$$

Draw this function. On the same graph sketch the value of the contract at various times up to expiry. What is the behaviour of the contract for large r?

10. Collect time-series data for the spot rate and the yield curve, either from back issues of newspapers or by writing to a bank. Using this data, plot a scatter diagram of the yield curve slope at the short end against the spot rate. Common models for the spot rate have the slope and spot rate related by (17.24). Is this borne out by your results?

11. Another interest rate model (Cox, Ingersoll & Ross 1990) has

$$u - \lambda w = ar^2$$

and

$$w = br^{3/2}.$$

Write down the zero-coupon bond pricing equation. Does this equation have any special properties?

18 Convertible Bonds

18.1 Introduction

In this final chapter, we bring together the models for asset prices and interest rates, in order to model securities that depend on both. For simplicity, we concentrate on the valuation of convertible bonds, although the ideas can easily be applied elsewhere. With the assumption of deterministic interest rates, these bonds are very similar to American vanilla options. We first illustrate the ideas with constant interest rates and, at the end of the chapter, we briefly bring together convertible bonds and stochastic interest rates in a two-factor model.

18.2 Convertible Bonds

A **convertible bond** has many of the same characteristics as an ordinary bond but with the additional feature that the bond may, at any time of the owner's choosing, be exchanged for a specified asset. This exchange is called **conversion**. The convertible bond on an underlying asset (with price S) returns Z, say, at time T *unless* at some previous time the owner has converted the bond into n of the underlying asset.[1] The bond may also pay a coupon to the holder.

Since the bond price depends on the value of that asset we have

$$V = V(S, t);$$

the contract value now depends on an asset price. It also depends on the time to maturity, but we usually suppress this dependence. Repeating

[1] We have implicitly assumed that the number of assets controlled by all the existing convertible bonds is small and that conversion does not affect the value of the issuing company. For further details see the Technical Point below.

the Black–Scholes analysis, with a portfolio consisting of one convertible bond and $-\Delta$ assets, we find that the change in the value of the portfolio is

$$d\Pi = \frac{\partial V}{\partial t}dt + \frac{\partial V}{\partial S}dS + \tfrac{1}{2}\sigma^2 S^2 \frac{\partial^2 V}{\partial S^2}dt - \Delta\, dS + K(S,t)\, dt$$

where we have included a coupon payment of $K(S,t)\,dt$ on the bond. As before, we choose

$$\Delta = \frac{\partial V}{\partial S}$$

to eliminate risk from this portfolio.

The return on this risk-free portfolio is at most that from a bank deposit and so

$$\frac{\partial V}{\partial t} + \tfrac{1}{2}\sigma^2 S^2 \frac{\partial^2 V}{\partial S^2} + (rS - D(S,t))\frac{\partial V}{\partial S} - rV + K(S,t) \le 0$$

for the bond price. This inequality is recognised as the basic Black–Scholes inequality but with the addition of the coupon payment term. The final condition is

$$V(S,T) = Z.$$

Recalling that the bond may be converted into n assets we have the constraint

$$V \ge nS.$$

In addition to this constraint, we require the continuity of V and $\partial V/\partial S$.

Thus the convertible bond is similar to an American option problem. It is interesting to note that the final data itself does not satisfy the pricing constraint. Thus, although the value *at* maturity may be Z the value *just before* is

$$\max(nS, Z).$$

Boundary conditions are

$$V(S,t) \sim nS \quad \text{as} \quad S \to \infty$$

and

$$V(0,t) = Ze^{-r(T-t)};$$

this last condition assumes (as can be verified *a posteriori*) that it is not optimal to exercise when $S = 0$.

This problem is as easy to solve numerically as an American option problem. It can be shown that an increase in D (respectively K) makes

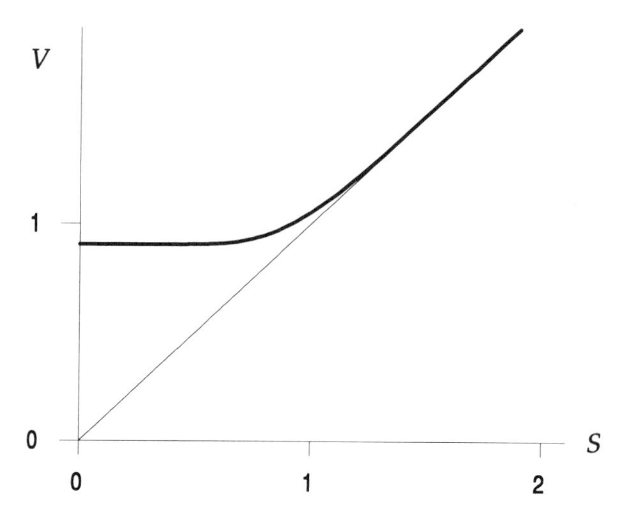

Figure 18.1 The value of a convertible bond with constant interest rate. See text for details.

early exercise more (less) likely. When $K = D = 0$ the constraint comes into play only at expiry and the convertible bond can be valued explicitly as a combination of cash and a European call option. In Figures 18.1 and 18.2 are shown the values of convertible bonds with $Z = 1$, $n = 1$, $r = 0.1$, $\sigma = 0.25$ and with one year before maturity. In both cases there are no coupon payments. In Figure 18.1 there is no dividend paid on the underlying but in Figure 18.2 we have $D_0 = 0.05$. Thus in the latter case there is a free boundary: for sufficiently large S the bond should be converted.

Sometimes the bond may be converted only during specified periods. This is called **intermittent conversion**. If this is the case then the constraint needs to be satisfied only during these times; at other times the contract is European.

18.2.1 Call and Put Features

The convertible bond permits the holder to swap the bond for a certain number of the underlying asset at any time of his choosing. Some bonds also have a **call feature**, which gives the company the right to purchase back the bond at any time (or during specified periods) for a fixed sum. Thus the bond with a call feature is worth less than the

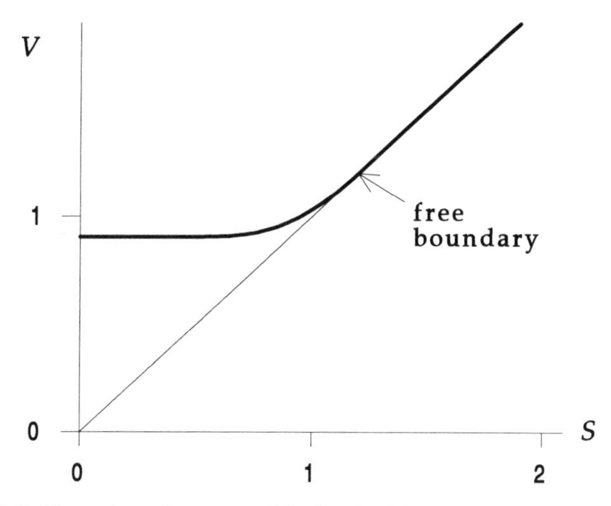

Figure 18.2 The value of a convertible bond with constant interest rate and a dividend paid on the underlying. See text for details.

bond without, since it cedes a right to the company. Such a call feature is easily modelled.

Suppose that the bond can be repurchased by the company for an amount M_1. The elimination of arbitrage opportunities leads to

$$V(S,t) \leq M_1.$$

Thus we must solve a constrained problem in which our bond price is bounded below by nS and above by M_1. (If $nS > M_1$, the bond should have been either converted or called, so it should not exist.) Again, V and $\partial V/\partial S$ must be continuous. As with the intermittent conversion feature it is also simple to incorporate the intermittent call feature, according to which the company can repurchase the bond only during certain time periods.

Some convertible bonds incorporate a **put feature**. This is a further right belonging to the owner of the bond. It allows the holder to return the bond to the issuing company for an amount M_2, say. Now we must impose the constraint

$$V(S,t) \geq M_2.$$

This feature increases the value of the bond.

18.3 Convertible Bonds with Random Interest Rate

When interest rates are stochastic, a convertible bond has a value of the form

$$V = V(S, r, t),$$

with the dependence on T suppressed. The value of the convertible bond is now a function of both S and r as independent variables (previously, r was just a parameter).

We assume that the asset price is governed by the standard model

$$dS = \sigma S\, dX_1 + \mu S\, dt, \tag{18.1}$$

and the interest rate by

$$dr = w(r, t)\, dX_2 + u(r, t)\, dt. \tag{18.2}$$

Since we are only modelling the bond, and do not intend to find explicit solutions, we allow u and w to be any functions of r and t. Observe that in (18.1) and (18.2) the Wiener processes have been given subscripts. This is because we are allowing S and r to be governed by two different random variables; this is a **two-factor model**. Thus, although dX_1 and dX_2 are both drawn from normal distributions with zero mean and variance dt, they are not necessarily the same random variable. They may, however, be correlated and we assume that

$$\mathcal{E}[dX_1\, dX_2] = \rho\, dt,$$

with $-1 \leq \rho(r, S, t) \leq 1$. We can still think of (18.1) and (18.2) as recipes for generating random walks for S and r, but now at each time-step we must draw two random numbers.

This is our first (and only) experience in this book of a **two-factor model**, in which there are two sources of risk and hence two independent asset-like variables in addition to t. In order to manipulate $V(S, r, t)$ we need to know how Itô's lemma applies to functions of two random variables. As might be expected, the usual Taylor series expansion together with a few rules of thumb result in the correct expression for the small change in any function of both S and r (see Exercise 5 of Chapter 2).

These rules of thumb are:

- $dX_1^2 = dt$;
- $dX_2^2 = dt$;
- $dX_1\, dX_2 = \rho\, dt$.

Applying Taylor's theorem to $V(S+dS, r+dr, t+dt)$ we find that

$$dV = \frac{\partial V}{\partial t}dt + \frac{\partial V}{\partial S}dS + \frac{\partial V}{\partial r}dr + \frac{1}{2}\left(\frac{\partial^2 V}{\partial S^2}dS^2 + 2\frac{\partial^2 V}{\partial S \partial r}dS\,dr + \frac{\partial^2 V}{\partial r^2}dr^2\right) + \cdots.$$

To leading order,

$$dS^2 = \sigma^2 S^2 dX_1^2 = \sigma^2 S^2 dt,$$

$$dr^2 = w^2 dX_2^2 = w^2 dt$$

and

$$dS\,dr = \sigma S w\,dX_1\,dX_2 = \rho\sigma S w\,dt.$$

Thus Itô's lemma for the two random variables governed by (18.1) and (18.2) becomes

$$dV = \frac{\partial V}{\partial t}dt + \frac{\partial V}{\partial S}dS + \frac{\partial V}{\partial r}dr + \frac{1}{2}\left(\sigma^2 S^2 \frac{\partial^2 V}{\partial S^2} + 2\rho\sigma S w \frac{\partial^2 V}{\partial S \partial r} + w^2 \frac{\partial^2 V}{\partial r^2}\right)dt. \tag{18.3}$$

Now we come to the pricing of the convertible bond. Let us construct a portfolio consisting of one bond with maturity T_1, $-\Delta_2$ bonds with maturity date T_2 and $-\Delta_1$ of the underlying asset. Thus

$$\Pi = V_1 - \Delta_2 V_2 - \Delta_1 S.$$

The analysis is much as before; the choice

$$\Delta_2 = \frac{\partial V_1/\partial r}{\partial V_2/\partial r}$$

and

$$\Delta_1 = \frac{\partial V_1}{\partial S} - \Delta_2 \frac{\partial V_2}{\partial S}$$

eliminates risk from the portfolio. Terms involving T_1 and T_2 may be grouped together separately to find that, dropping the subscripts,

$$\frac{\partial V}{\partial t} + \frac{1}{2}\left(\sigma^2 S^2 \frac{\partial^2 V}{\partial S^2} + 2\rho\sigma S w \frac{\partial^2 V}{\partial S \partial r} + w^2 \frac{\partial^2 V}{\partial r^2}\right) + rS\frac{\partial V}{\partial S} + (u - w\lambda)\frac{\partial V}{\partial r} - rV = 0,$$

where again $\lambda(r, S, t)$ is the market price of interest rate risk.

This is the convertible bond pricing equation. Note that it contains the known interest rate problem ($u = 0 = w$, i.e. Black–Scholes) and the simple bond problem ($\partial/\partial S = 0$) as special cases. More generally, when the underlying asset pays dividends and the bond pays a coupon we have

$$\frac{\partial V}{\partial t} + \frac{1}{2}\left(\sigma^2 S^2 \frac{\partial^2 V}{\partial S^2} + 2\rho\sigma S w \frac{\partial^2 V}{\partial S \partial r} + w^2 \frac{\partial^2 V}{\partial r^2}\right) + (rS - D)\frac{\partial V}{\partial S} + (u - w\lambda)\frac{\partial V}{\partial r} - rV + K = 0.$$

The final condition and American constraints are exactly as before; there is one constraint each for the convertibility feature and the call feature. Since this is a diffusion equation with two 'space-like' state variables S and r – that is, there are second derivatives of V with respect to each of S and r, as well as a cross-term – we need to impose boundary conditions on the limits of the (S, r) space. In other words, we must prescribe $V(0, r, t)$ and $V(\infty, r, t)$ for all t, $V(S, \infty, t)$ for all S and t and a second boundary condition on a fixed r boundary, again for all S and t. Some of these boundary conditions are obvious and others, less obviously, are a result of insisting that V remain finite. For example, for a convertible bond with no call feature we have

$$V(S, r, t) \sim nS \quad \text{as} \quad S \to \infty;$$

$V(0, r, t)$ is given by the solution of the simple bond problem (no convertibility and stochastic interest rates);

$$V(S, r, t) \to 0 \quad \text{as} \quad r \to \infty;$$

and the last boundary condition, to be applied on the lower r boundary, is equivalent to boundedness of V.

Technical Point: The Issue of New Shares.
We have assumed in this chapter that the existence of convertible bonds does not affect the market worth of the company. In reality, the conversion of the bond into n shares requires the company to issue n new shares. This contrasts with options for which the exercise leaves the number of shares unchanged. We do not include any of the details here; however, if we let S be the worth of the company's assets, without the bond obligation, and N be the number of shares before conversion, we arrive at the following problem.

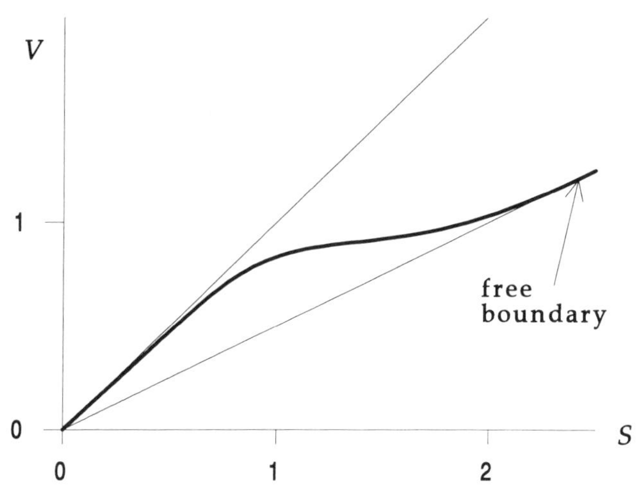

Figure 18.3 The value of a convertible bond versus company's assets, allowing for dilution on issue of new shares.

The convertible bond pricing equation (with known or stochastic interest rate) is solved with the constraints

$$V \geq \frac{nS}{n+N}, \tag{18.4}$$

$$V \leq S \tag{18.5}$$

and

$$V(S,T) = Z.$$

Constraint (18.4) bounds the bond price below by its value on conversion, and constraint (18.5) allows the company to declare bankruptcy if the bond becomes too valuable. The factor $N/(n+N)$ is known as the dilution. A typical convertible bond value is shown in Figure 18.3. In this example we have $Z = 1$, $r = 0.1$, $\sigma = 0.25$, $D_0 = 0.05$, there is one year to maturity and the dilution factor is 0.5. In the limit $n/N \to 0$ this model is identical to the one considered above.

Note that the total worth of the company is $S - V$ and the share price is thus $(S - V)/N$ and *not* S.

Further Reading

- For details of the effect of the issue of new shares on the value of convertible bonds see Brennan & Schwartz (1977), Cox & Rubinstein (1985) and Gemmill (1992).

Exercises

1. Assuming that interest rates are constant, find an explicit formula for the value of a convertible bond with zero coupon and zero dividend on the underlying and generalise this result to include a coupon payment.

2. Perform a local analysis of the same kind as those in Chapter 7 to find the position of the free boundary close to maturity. Allow for both a dividend payment on the underlying and a coupon payment on the bond.

3. For an arbitrary interest rate model (18.3) must be solved numerically in three dimensions. Are there any special interest rate models that allow a similarity reduction to two dimensions? Do explicit formulæ exist for the value of the convertible bond?

Hints to Selected Exercises

Chapter 1

1. Since the holder of one share just before a one-for-one split will hold two just after, and since the value and future prospects of the company are unaffected by what is only a nominal change, the stock price must be halved on the introduction of the new shares. The exercise prices of options are therefore also all halved, as are their values.
2. $S - D$; see Chapter 6.
3. Yes. One reason is that options provide insurance, and the larger the volatility the more the need for insurance against an unfavourable outcome from a position in the stock, and consequently the greater the expense. On the other side of the coin, when the volatility is high, the asset price is more likely to be high or low when the option expires. High prices make calls more valuable, low prices favour puts. We show these results in more detail later (see Chapter 5).

Chapter 2

1. Use equation (2.7).
3. Use the cumulative distribution function and the fact that for any $s > 0$, $\text{prob}[S < s] = \text{prob}[\log S < \log s]$; then differentiate to find the density function of S.
4. Use the general version of Itô's lemma given on page 27 to find a differential equation for $f(G)$.
5. As with a function of a single variable, use Taylor's theorem with the additional 'rules'

$$dX_i^2 \to dt \quad \text{and} \quad dX_i\, dX_j \to \rho_{ij}\, dt.$$

Chapter 3

4. Remember that $N(-\infty) = 0$ and $N(\infty) = 1$.

5. The Δ is the slope of the graphs of the option values. Consider a written call and a realisation that rises, with $S > E$ at expiry. The option is exercised and the writer must deliver the asset, receiving E in return. This money, together with the initial premium (allowing for interest rates), exactly balances the cost of the initial hedge and subsequent purchases of the asset (some of which are at a price below the final value of S). The other cases are similar.

6. (a) The equation is linear and homogeneous, so try powers of S. (b) This separation of variables gives

$$\frac{A'}{A} = -\left(\tfrac{1}{2}\sigma^2 S^2 B'' + rSB' - rB\right)/\,B;$$

since the left-hand side is a function of t only, and the right-hand side is a function of S only, both must be equal to a constant. The two resulting ordinary differential equations are easy to solve (see part (a) for $B(S)$).

7. (a) Long one share and short one call cannot make a loss. (b) Buy one share and write one call. At expiry this portfolio is worth at most E, so now it is worth at most the present value of E. (c) If the first inequality is not true, buy the call with exercise E_1 and write the other. If the second is not true, write the call with exercise E_1 and buy the other. The greatest possible loss at expiry is less than the result of investing the profit from the initial transaction until expiry (Do you see how to sharpen this result slightly?) (d) If the result is not true, buy the longer dated call and write the other.

8. Assume the option *will* be exercised. The writer can deliver the asset at exercise by buying it now. How much will this strategy cost?

9. The portfolio is $\Pi = C - \Delta S$. If $dS = +1$, $C = 101 - 100 = 1$ at expiry and $dC = (1 - C)$, while if $dS = -1$, $C = 0$ at expiry and so $dC = -C$. Thus for an up move $d\Pi = 1 - C - \Delta$ and for a down move $d\Pi = 0 - C + \Delta$. The choice $\Delta = \frac{1}{2}$ gives $d\Pi = 0$ (since $r = 0$) and $C = \frac{1}{2}$ (i.e. 50¢). Notice that p does not feature in the solution. The reader should now repeat the calculation for the cash-or-nothing option.

Chapter 4

2. Write $u(x, 0) = \sum_1^\infty a_n \sin nx$, where a_n is given by the usual Fourier integral. Then $u(x, \tau) = \sum_1^\infty a_n e^{\mp n^2 \tau} \sin nx$ respectively; the $+$ sign in the exponent, corresponding to the backward equation, almost always gives finite-time divergence in the series.

Chapter 5

1. Try $u(x,\tau) = f(x/\sqrt{\tau})$; you need $f(-\infty) = 0$ and $f(\infty) = 1$ to fit the boundary and initial conditions. Note that the derivative of the initial data for u is the delta function.

2. Since $v(0,\tau) = u(0,\tau)$ and $v(0,\tau) = -u(0,\tau)$, $v(0,\tau) = 0$. Now write the solution for $v(x,\tau)$ as an integral from $-\infty$ to ∞, split the range of integration into two, from $-\infty$ to 0 and 0 to ∞, and change variable in the former. The second term in the integral can also be found by replacing x by $-x$ in the diffusion equation, a change of variable that leaves it invariant.

3. (a) Try solutions of the form $\tau^\alpha f(x/\sqrt{\tau})$ to find that $\alpha = \frac{3}{2}$. (b) Try $\tau f(x/\sqrt{\tau})$.

4. The first part is an extension of the argument on page 77. For the second, try a change of variables $\hat{\tau} = F(\tau)$ to get to the heat equation. Then use the usual Black–Scholes formulæ.

9. For a call,

$$\Gamma = \partial\Delta/\partial S = N'(d_1)/\sigma S\sqrt{T-t};$$

$$\Theta = \partial C/\partial t = -\tfrac{1}{2}\sigma^2 S^2\Gamma - rS\Delta + rC$$

from the Black–Scholes equation, which gives

$$\Theta = -\tfrac{1}{2}\sigma S N'(d_1)/\sqrt{T-t} - rEe^{-r(T-t)}N(d_2).$$

The call vega and rho are

$$\partial C/\partial\sigma = SN'(d_1)\sqrt{T-t}, \qquad \partial C/\partial r = E(T-t)e^{-r(T-t)}N(d_2).$$

The corresponding results for puts can most easily be obtained from put-call parity.

11. Itô's lemma gives

$$dC = \frac{\partial C}{\partial S}dS + \left(\frac{\partial C}{\partial t} + \tfrac{1}{2}\sigma^2 S^2\frac{\partial^2 C}{\partial S^2}\right)dt.$$

The Black–Scholes equation can be used to simplify the second term.

12. Use equation (5.7).

13. Write down the Black–Scholes equations for the two calls. In the equation for C_1, rewrite σ_1 in the form $\sigma_1 - \sigma_2 + \sigma_2$, and subtract the equation for C_2 from the result. Now relate $C_1 - C_2$ to $u(x,\tau)$ of the question, with all terms involving only C_1 becoming $f(x,\tau)$.

15. Note that $\mathcal{H}(E-S)+\mathcal{H}(S-E)=1$.

17. Integrate (2.10) from E to ∞ (replace S by the integration variable S', and S_0 by S).

18. It is sufficient to synthesise the payoff $\Lambda(S)$. This gives

$$\Lambda(S)=\int_0^\infty f(E)\max(S-E,0)\,dE=\int_0^S f(E)(S-E)\,dE.$$

Differentiating twice, $f(S)=\Lambda''(S)$. This may also be found by writing $\max(S-E,0)=(S-E)\mathcal{H}(S-E)$ and noting that the first S-derivative of this is $\mathcal{H}(S-E)$ and so the second is $\delta(S-E)$; we have used the result that $(S-E)\delta(S-E)=0$. When $\Lambda(S)=\max(S-E,0)$, $\Lambda''(S)=\delta(S-E)$, corresponding to one call with exercise price E. The case $\Lambda(S)=S$ should really be written $\Lambda(S)=\max(S,0)$, and so the synthesizing portfolio is just S (which is of course a call with exercise price zero!). For the binary call we get, in a similar way, $f(S)=\Lambda'(S)$.

19. We can write $C=EF(S/E)$ for some function F (in fact it is given by the Black–Scholes formulæ). Then

$$\frac{\partial C}{\partial E}=C/E-(S/E)F'(S/E)$$

and since $\partial C/\partial S=F'(S/E)$, $S\partial C/\partial S=C-E\partial C/\partial E$. Similarly, $S^2\partial^2 C/\partial S^2=E^2\partial C/\partial E^2$; substitution into the Black–Scholes equation gives the result.

20. Can there be two risk-free assets with different rates of return?

Chapter 6

1. Solve the Black–Scholes equation for $C-P$ (with dividends) using the final data $C-P=S-E$.

2. $\Delta=e^{-D_0(T-t)}N(d_{10})$.

3. Follow the procedure of Chapter 5. The transformations for S and t are as before, and

$$V(S,t)=Ee^{-\frac{1}{2}(k'-1)x-\left(\frac{1}{4}(k'-1)^2+k\right)\tau}u(x,\tau),$$

where $k=r/\frac{1}{2}\sigma^2$, also as before, and $k'=(r-D_0)/\frac{1}{2}\sigma^2$. There are now three dimensionless parameters: k, k' and $\frac{1}{2}\sigma^2T$. The payoff for a call is $\max(e^{\frac{1}{2}(k'+1)x}-e^{\frac{1}{2}(k'-1)x},0)$.

4. As $S\to\infty$, $C\sim Se^{-D_0(T-t)}$, which for large enough S is certainly below the payoff if $D_0 > 0$. (This asymptotic behaviour can be seen by balancing the first order and undifferentiated terms in the Black–Scholes equation.) For the second part, take the term $D_0 S\,\partial C/\partial S$ onto the other side of the equation and note that it is positive.

5. Follow the argument in the text. The put is more valuable with dividends, because they decrease the value of the stock.

6. Follow the argument in the text but over two stages.

7. S must fall by an amount D, so $V(S, t_d^-) = V(S - D, t_d^+)$. What would happen if $S = 1$ and $D = 1.20$? A situation such as this might arise if S fell suddenly between the announcement of a dividend and the date it is paid.

8. The party who is short the contract receives Z at initiation, so need only borrow $S - Z$ at a cost of $(S - Z)e^{r(T-t)}$, which is therefore the forward price in this case.

9. Since $F = Se^{r(T-t)}$,

$$\begin{aligned} dF &= dSe^{r(T-t)} - rSe^{r(T-t)}\,dt \\ &= (\sigma S\,dX + \mu S\,dt)e^{r(T-t)} - rF\,dt \\ &= \sigma F\,dX + (\mu - r)F\,dt. \end{aligned}$$

Where did we use Itô's lemma in the above?

11. Follow the procedure for an option on such an asset.

12. Work out the payoff and decompose the contract into options.

13. The third term in equations (6.15) and (6.16) have $r(t)$ replaced by $r(t) - D(t)$, as does the integrand in the expression for $\alpha(t)$.

Chapter 7

1. Let the instalment option have value $V(S,t)$, and $\Pi = V - \Delta S$ be the usual Black–Scholes portfolio (from the point of view of the holder of the option). When we work out $d\Pi$, we must subtract $L\,dt$ to account for the instalment payment before equating the result to $r\Pi\,dt$. Since the option should always have a positive value (if not, the holder just stops paying the instalments and the option lapses), we impose $V \geq 0$. Where $V > 0$, it satisfies $\mathcal{L}_{BS}V = L$, so in linear complementarity form we have

$$V \cdot (\mathcal{L}_{BS}V - L) = 0, \quad V \geq 0, \quad \mathcal{L}_{BS}V \leq 0,$$

and V and $\partial V/\partial S$ are continuous. For a call there is just one free boundary, and $V = 0$ for values of S below this.

2. The first inequality says the put must be above its payoff. For the second, consider the portfolio $C - S + Ee^{-r(T-t)}$ (long a call, short a share, with a bank deposit that will equal E at expiry). If we exercise the option at any time before expiry, paying E and receiving S, the result is negative. But at expiry the portfolio has zero value if $S \geq E$ and positive value if $S < E$. It is therefore not optimal to exercise the option before expiry, since by waiting we can obtain a better outcome. Therefore the American call (without dividends) has the same value as its European counterpart, while the American put is more valuable than the European put. The second part of the second inequality follows immediately from this statement and put-call parity for European options. The first part follows by considering $C - P + E - S$. The call will never be exercised. If the put is exercised early the result is positive, while if the portfolio is left to expiry it is worthless. (Note that we have ignored the effect of interest rates here.)

3. By symmetry we need only find one free boundary, say $x = x_f > 0$. For $x > x_f$, the equation of the string is that of the straight line joining $(x_f, \frac{1}{2} - x_f^2)$ to $(1, 0)$. The slope of this line, $-(\frac{1}{2} - x_f)/(1 - x_f)$, must be equal to that of the obstacle at the free boundary, $-x_f$. This soon gives $x_f = 1 - 1/\sqrt{2}$.

4. Expanding for small λ,

$$E[(1-\lambda)u + \lambda v] - E[u] = \lambda \int_{-1}^{1} u'(v-u)'\,dx + O(\lambda^2).$$

By taking λ very small, we see that the coefficient of λ must be positive.

5. Use equations (7.19) and (7.20).

6. Choose $X(x) = e^{\frac{1}{2}(k+1)x}$; then $w(x, \tau)$ satisfies the diffusion equation for $-\infty < x < 0$ with initial data $w(x, 0) = -e^{\frac{1}{2}(k+1)x}$ and with $w(0, \tau) = 0$. Solve by images (see the discussion in Chapter 12) to get

$$\begin{aligned} w(x,\tau) &= e^{\frac{1}{4}(k+1)^2\tau}\left(e^{-\frac{1}{2}(k+1)x} N\left(\frac{x-(k+1)\tau}{\sqrt{2\tau}}\right)\right. \\ &\qquad \left. - e^{\frac{1}{2}(k+1)x} N\left(\frac{-x-(k+1)\tau}{\sqrt{2\tau}}\right)\right). \end{aligned}$$

Going back into financial variables, we get

$$V(S,t) = (E/S)^{r/\frac{1}{2}\sigma^2} N(d_4) + (S/E)N(d_1),$$

where d_1 is as usual and

$$d_4 = \frac{\log(S/E) - (r + \frac{1}{2}\sigma^2)(T-t)}{\sigma\sqrt{(T-t)}}.$$

7. For any American option with payoff $\Lambda(S)$,

$$V \geq \Lambda, \quad \mathcal{L}_{BS}V \leq 0, \quad (V - \Lambda) \cdot \mathcal{L}_{BS}V = 0.$$

8. With $\xi = x/\sqrt{t}$, $u^*(\xi)$ satisfies $(u^*)'' + \frac{1}{2}\xi(u^*)' = 0$, with $u^*(0) = 1$, $u^*(\xi_0) = 0$, $(u^*)'(\xi_0) = -\frac{1}{2}\xi_0$. The solution with the first two of these boundary conditions is

$$u(\xi) = 1 - \int_0^{\xi} e^{-\frac{1}{4}s^2}\,ds \Big/ \int_0^{\xi_0} e^{-\frac{1}{4}s^2}\,ds\,,$$

and the third boundary condition gives the transcendental equation.
9. The working is almost identical to that of Exercise 8.

Chapter 8

4. Write the algorithm as $u_n^{m+1} = (1 - 2\alpha)u_n^m + \alpha(u_{n+1}^m + u_{n-1}^m)$. This requires $2N$ multiplications and/or divisions per time-step.
5. The e_n^m satisfy (8.10) because it is linear. Direct substitution of the suggested form leads to

$$\lambda = 1 + \alpha\left(\frac{\sin(n+1)\omega - 2\sin n\omega + \sin(n-1)\omega}{\sin n\omega}\right).$$

Now use the identity $\sin(A + B) + \sin(A - B) = 2\sin A\cos B$ to obtain the given expression. If $\alpha > \frac{1}{2}$ we see that for some values of ω we will have $|\lambda| > 1$, in which case $|e_n^m| \to \infty$ as $m \to \infty$. This is instability.
7. The only change is to the payoff function.
9. The method is explicit because we solve the Black–Scholes equation backward in time. We start with the payoff and calculate V_n^m from V_n^{m+1} and $V_{n\pm1}^{m+1}$. To see the sort of stability problems that could arise, regard δt as given and consider what happens to b_n as $n \to \infty$.
14. Write

$$\begin{pmatrix} B_0 & C_0 & 0 & \cdots & 0 \\ A_1 & B_1 & C_1 & & \vdots \\ 0 & A_2 & \ddots & \ddots & 0 \\ \vdots & & \ddots & \ddots & C_{N-1} \\ 0 & \cdots & 0 & A_N & B_N \end{pmatrix} =$$

$$\begin{pmatrix} 1 & 0 & 0 & \cdots & 0 \\ D_1 & 1 & 0 & & \vdots \\ 0 & D_2 & \ddots & \ddots & 0 \\ \vdots & & \ddots & \ddots & 0 \\ 0 & \cdots & 0 & D_N & 1 \end{pmatrix} \times \begin{pmatrix} F_0 & E_0 & 0 & \cdots & 0 \\ 0 & F_1 & E_1 & & \vdots \\ 0 & 0 & \ddots & \ddots & 0 \\ \vdots & & \ddots & \ddots & E_{N-1} \\ 0 & \cdots & \cdots & 0 & F_N \end{pmatrix},$$

then eliminate the D_n and E_n. Let $\mathbf{L}$ and $\mathbf{U}$ be the lower triangular and upper triangular matrices on the right-hand side of this equation. Write $\mathbf{L}$ and $\mathbf{U}$ in terms of A_n, B_n , C_n and F_n, then solve the systems $\mathbf{L}\boldsymbol{q}^m = \boldsymbol{V}^{m+1}$ by forward substitution and $\mathbf{U}\boldsymbol{V}^m = \boldsymbol{q}^m$ by backward substitution.

15. Starting with a guess $V_n^{m,0}$, Jacobi is

$$V_n^{m,k+1} = \frac{1}{B_n}\left(V_n^{m+1} - A_n V_{n-1}^{m,k} - C_n V_{n+1}^{m,k}\right),$$

Gauss–Seidel is

$$V_n^{m,k+1} = \frac{1}{B_n}\left(V_n^{m+1} - A_n V_{n-1}^{m,k+1} - C_n V_{n+1}^{m,k}\right)$$

and SOR is

$$\begin{aligned} y_n^{m,k+1} &= \frac{1}{B_n}\left(V_n^{m+1} - A_n V_{n-1}^{m,k+1} - C_n V_{n+1}^{m,k}\right), \\ V_n^{m,k+1} &= V_n^{m,k} + \omega(y_n^{m,k+1} - V_n^{m,k}). \end{aligned}$$

18. Average the explicit method described in Exercise 9 and the implicit method described in Exercise 13; this gives the Crank–Nicolson method. The LU decomposition solution is essentially the same as described in Exercise 14, with minor modifications to the values of A_n, B_n and C_n as well as the right-hand side of the system of equations. Similarly, the Jacobi, Gauss–Seidel and SOR methods are essentially the same as those of Exercise 15 with adjustments to the values of A_n, B_n and C_n.

19. The explicit method is

$$u_{ij}^{m+1} = (1-4\alpha)u_{ij}^m + \alpha\left(u_{i+1\,j}^m + u_{i-1\,j}^m + u_{i\,j+1}^m + u_{i\,j-1}^m\right)$$

and is stable for $0 < \alpha \le \frac{1}{4}$. The fully implicit method is

$$(1+4\alpha)u_{ij}^m - \alpha\left(u_{i+1\,j}^m + u_{i-1\,j}^m + u_{i\,j+1}^m + u_{i\,j-1}^m\right) = u_{i,j}^{m-1}$$

and the Crank–Nicolson method is

$$\begin{aligned} (1+2\alpha)u_{ij}^m &- \tfrac{1}{2}\alpha\left(u_{i+1\,j}^m + u_{i-1\,j}^m + u_{i\,j+1}^m + u_{i\,j-1}^m\right) \\ &= (1-2\alpha)u_{i,j}^{m-1} + \tfrac{1}{2}\alpha\left(u_{i+1\,j}^{m-1} + u_{i-1\,j}^{m-1} + u_{i\,j+1}^{m-1} + u_{i\,j-1}^{m-1}\right). \end{aligned}$$

The latter two are stable for all $\alpha > 0$. The values at the boundary of the grid are given from

$$u^m_{N^+\,j} = u^1_{\infty}(j\,\delta y, m\,\delta\tau), \quad u^m_{N^-\,j} = u^1_{-\infty}(j\,\delta y, m\,\delta\tau),$$
$$u^m_{i,N^+} = u^2_{\infty}(i\,\delta x, m\,\delta\tau), \quad u^m_{i,N^-} = u^2_{-\infty}(i\,\delta x, m\,\delta\tau).$$

Chapter 9

2. The explicit finite-difference inequalities become

$$u^{m+1}_n \geq (1-2\alpha)u^m_n + \alpha(u^m_{n-1} + u^m_{n+1}), \quad u^{m+1}_n \geq g^{m+1}_n.$$

The finite-difference linear complementarity form can be written as

$$\mathbf{I}\boldsymbol{u}^{m+1} = \boldsymbol{Z}^m, \quad \boldsymbol{u}^{m+1} \geq \boldsymbol{g}^{m+1}, \quad (\boldsymbol{u}^{m+1} - \boldsymbol{g}^{m+1}) \cdot (\boldsymbol{u}^{m+1} - \boldsymbol{Z}^m) = 0.$$

Applying the SOR algorithm with the identity matrix **I** gives the stated algorithm.

3. This is essentially the same as the Crank–Nicolson formulation described in the text.

4. This is essentially the same as the Crank–Nicolson formulation described in the text; replace $\frac{1}{2}\alpha$ by α and Z^m_n by u^m_n.

5. See Exercise 9 of the previous chapter to obtain the explicit finite-difference inequality, and then Exercise 1 of this chapter to see how to solve the approximate linear complementarity problem. The only change necessary for an American cash-or-nothing call would be a change in the payoff for early exercise and/or exercise at expiry.

5. See Exercises 13 and 15 of the previous chapter. The linear complementarity formulation follows as it does in the text.

7. This is

$$y^{m,k+1}_n = \frac{1}{B_n}\left(V^{m+1}_n - A_n V^{m,k+1}_{n-1} - B_n V^{m,k}_{n+1}\right),$$
$$V^{m,k+1}_n = \max\left(\max(E - n\,\delta S), V^{m,k}_n + \omega\left(y^{m,k+1}_n + V^{m,k+1}_n\right)\right).$$

Chapter 11

1. What is the final value of the portfolio that is long a call-on-a-call and short a put-on-a-call?

2. It is not easy to compute the random walk for C *in terms of* C *itself* rather than in terms of S.

3. The answer is

$$V(S,t) = SM(a_+, b_+; \rho) - E_2 e^{r(T_2-t)} M(a_-, b_-; \rho) - E_1 e^{-r(T_1-t)} N(a_-),$$

where

$$a_\pm = \frac{\log(S/S_1) + (r \pm \frac{1}{2}\sigma^2)(T_1 - t)}{\sigma\sqrt{T_1 - t}},$$

$$b_\pm = \frac{\log(S/E_1) + (r \pm \frac{1}{2}\sigma^2)(T_2 - t)}{\sigma\sqrt{T_2 - t}},$$

$$\rho = \sqrt{\frac{T_1 - t}{T_2 - t}},$$

and $M(a, b; \rho)$ is the cumulative distribution function of the bivariate normal distribution with correlation coefficient ρ; lastly S_1 is the value of S at which $C_{BS}(S, T_1) = E_1$. Might it not be easier to value this product numerically?
5. What is the payoff?
6. Use put-call parity to substitute for the put in terms of the call. The result is a package of vanilla options.

Chapter 12

1. The out boundary condition becomes $V(X,t) = Z$.
2. The in final condition becomes $V(S,T) = Z$.
4. For the down-and-out call, the delta is positive for $S > X$ and zero for $S \leq X$. For the down-and-in call, the delta is negative for $S > X$ (so this option looks more like a put) and switches to a positive value, the delta for a vanilla call, at $S = X$. If the asset reaches the barrier, the cost of the hedge for the underlying call is financed by short selling as S decreases to X, as well as by the premium.
5. Expand the final value in a Fourier series (after transforming to the diffusion equation). See Exercise 2 of Chapter 4.
6. Solve the ordinary differential equation

$$\tfrac{1}{2}\sigma^2 S^2 \frac{d^2V}{dS^2} + rS\frac{dV}{dS} - rV = 0$$

for $V(S)$, with the appropriate boundary conditions at the barrier and at $S = 0$ or as $S \to \infty$ (whichever is appropriate).
7. Ask the question: what contract do I receive if the asset reaches the first barrier? Then use that as a boundary condition on that barrier.

9. It is a property of random walks that if they reach a certain level they will cross that level infinitely often. This is a result of the scaling of the Wiener process with the square root of time. A better definition of a bouncing ball option would specify time periods, say of one week's duration, and if the random walk reaches the barrier in that period it counts only as one hit.

Chapter 13

1. The discrete running sum can be represented by

$$\int_0^t S(\tau) \sum_{i=1}^{N} \delta(\tau - t_i)\, d\tau.$$

This puts discrete sampling into our general framework with

$$f(S,t) = S \sum_{i=1}^{N} \delta(t - t_i).$$

Use this in the partial differential equation and consider what happens across a sampling date.
2. Do not be over-ambitious in inventing new options!

Chapter 14

3. Look for a solution that is linear in S and I.
4. See the next chapter.
5. Look for a solution of the form $V(S,I,t) = H(R,t)$ with an appropriate definition of I, and with R as in the text.

Chapter 15

1. See page 245.
2. All of these options have similarity solutions. In each case they have the form $V(S,J,t) = JH(\xi,t)$, where $\xi = S/J$. This reduces the dimensionality of the problem. The equation for $H(\xi,t)$ is simply the Black–Scholes equation, with $H(\xi,T)$ given by the payoff function and, on $\xi = 1$,

$$\frac{\partial H}{\partial \xi} + H = 0.$$

Chapter 16

1. Take costs of the form

$$\kappa_1 + \kappa_2|\nu| + \kappa_3|\nu|S.$$

Now subtract the expected value of this from the value of the hedged portfolio.
2. As an example, take $\kappa_2 = \kappa_3 = 0$ with $\kappa_1 > 0$. Find an explicit solution of the resulting equation. What constraint could you impose to improve this hedging strategy? What would this mean in practice?
3. Look for a similarity solution of the form

$$u(x,t) = t^{-1/2}h(x/\sqrt{t}).$$

There will be a different solution for $|x/\sqrt{t}| < \xi_0$ and $|x/\sqrt{t}| > \xi_0$ for some unknown ξ_0. Use smoothness conditions to find ξ_0.
4. Write $V(S,t) = V_0(S,t) + KV_1(S,t) + \cdots$. Find equations for V_0 and V_1 separately.
5. Let the value of the initial portfolio plus the new contract be $V(S,t) + \bar{V}(S,t)$. Find the equation and final condition satisfied by $\bar{V}(S,t)$ to leading order.

Chapter 17

4. Introduce Dirac delta functions into the equation by writing the exchange of payments in the form

$$(r - r^*)\sum \delta(t - t_i).$$

Show by both mathematical and financial arguments that this leads to jump conditions.
5. Look for a solution of the swap equation of the form

$$(A(t;T) + C(t;T)r)e^{-rB(t;T)}.$$

6. Write

$$V(r,t;T) = a(r) + b(r)(T-t) + c(r)(T-t)^2 + \cdots;$$

find a, b and c.
7. As for Exercise 6 but with different final data.
8. This approach assumes that bond prices follow a lognormal random walk. This may be a tolerable assumption in the short term and has

the advantage that there are simple formulæ for common options. In the long term, however, since the bond price is known at maturity, the lognormal random walk assumption is clearly a poor one.
11. Look for a similarity solution.

Chapter 18

1. Assume that the bond is a European option with payoff $\max(nS, Z)$. Write this in terms of cash plus a vanilla call option and show that the constraint $V \geq nS$ is always satisfied.

Bibliography

Babbs, S. 1992 Binomial valuation of lookback options. *Midland Montagu, working paper.*

Baker, C.T.H. (1977) *The Numerical Treatment of Integral Equations*, Clarendon Press, Oxford.

Barone-Adesi, G. & Whaley, R.E. (1987) Efficient analytic approximation of American option values. J. Fin. **42** 301–320.

Barrett, J.W., Moore, G. & Wilmott, P. (1992) Inelegant efficiency. *Risk* Magazine, **9** (5) 82–84.

Black, F., Derman, E. & Toy, W. (1990) A one-factor model of interest rates and its application to Treasury bond options. Finan. Analysts J. **46** 33–9.

Black, F. & Scholes, M. (1973) The pricing of options and corporate liabilities. J. Pol. Econ. **81** 637–659.

Blank, S.C., Carter, C.A. & Schmiesing, B.H. (1991) *Futures and Options Markets*, Prentice-Hall.

Boyle, P.P. (1991) Multi-asset path-dependent options. FORC Conf., Warwick.

Boyle, P.P. & Emanuel, D. (1980) Discretely adjusted option hedges. J. Finan. Econ. **8** 259–282.

Brealey, R.A. (1983) *An Introduction to Risk and Return from Common Stock*, MIT Press, Cambridge, Mass.

Brennan, M. & Schwartz, E. (1977) Convertible bonds: valuation and optimal strategies for call and conversion. J. Fin. **32** 1699–1715.

Brennan, M. & Schwartz, E. (1978) Finite-difference methods and jump processes arising in the pricing of contingent claims: a synthesis. J. Finan. Quant. An. **13** 462–474

Carslaw, H.S. & Jaeger, J.C. (1989) *Conduction of Heat in Solids*, Oxford University Press.

Conze, A. & Viswanathan (1991) Path-dependent options—the case of lookback options. J. Fin. **46** 1893–1907.

Copeland, T., Koller, T. & Murrin, J. (1990) *Valuation: measuring and managing the value of companies*, Wiley.

Cox, J.C., Ingersoll, J. & Ross, S. (1981) The relationship between forward prices and futures prices. J. Fin. Econ. **9** 321–346.

Cox, J.C., Ingersoll, J. & Ross, S. (1985) A theory of the term structure of interest rates. Econometrica **53** 385–467.

Cox, J.C. & Ross, S.A. (1976) The valuation of options for alternative stochastic processes. J. Fin. Econ. **3** 145–166.

Cox, J.C. & Rubinstein, M. (1985) *Options Markets*, Prentice Hall.

Crank, J., (1984) *Free and Moving Boundary Problems*, Oxford University Press.

Crank, J. (1989) *Mathematics of Diffusion*, Oxford University Press.

Cryer, C.W. (1971) The solution of a quadratic programming problem using systematic overrelaxation, SIAM J. Control **9** 385–392.

Davis, M.H.A. & Norman, A.R. (1990) Portfolio selection with transaction costs. Math. Oper. Res. **15** 676–713.

Davis, M.H.A., Panas, V.G. & Zariphopoulou, T. (1993) European option pricing with transaction costs. SIAM J. Control **31** 470–493.

Dewynne, J.N., Whalley, A.E. & Wilmott, P. (1994) Path-dependent options and transaction costs. Phil. Trans. R. Soc. Lond. **A347**, 517–529.

Dewynne, J.N. & Wilmott, P. (1993 a) Partial to the exotic. *Risk* Magazine **6** (3) 38–46.

Dewynne, J.N. & Wilmott, P. (1993 b) Lookback options. OCIAM working paper, Mathematical Institute, Oxford University.

Dewynne, J.N. & Wilmott, P. (1993 c) Average strike options—European and American. OCIAM working paper, Mathematical Institute, Oxford University.

Dewynne, J.N. & Wilmott, P. (1995) A note on average rate options with discrete sampling. SIAM J. Appl. Math., **55** 267–276.

Dothan, M.U. (1978) On the term structure of interest rates. J. Finan. Econ. **6** 59–69.

Duffie, D. (1992) *Dynamic Asset Pricing Theory*, Princeton.

Duffie, J.D. & Harrison, J.M. (1992) Arbitrage pricing of Russian options and perpetual lookback options. Working paper.

Elliott, C.M. & Ockendon, J.R. (1982) *Weak and Variational Methods for Free and Moving Boundary Problems*, Pitman.

Fama, E. (1965) The behaviour of stock prices. J. Bus. **38** 34–105.

Fitt, A.D., Dewynne, J.N. & Wilmott, P. (1994) An integral equation for the value of a stop-loss option. In *Proc. Seventh Euro. Conf. Mathematics in Industry* (eds. Fasano, A. & Primicerio, M.). B.G. Teubner.

Friedman, A. (1988) *Variational Principles and Free Boundary Problems*, Robert E. Krieger Publishing.

Geman, H. & Yor, M. (1992) Bessel processes, Asian options and perpetuities. FORC Conf., Warwick.

Gemmill, G. (1992) *Options Pricing*, McGraw-Hill.

Geske, R. (1978) Pricing of options with stochastic dividend yield. J. Fin. **33** 617–25.

Geske, R. (1979) The valuation of compound options. J. Finan. Econ. **7** 63–81.

Geske, R. & Shastri, K. (1985) Valuation by approximation: a comparison of alternative option valuation techniques, J. Fin. & Quant. Anal. **20**, 45–71.

Goldman, M.B., Sozin, H.B. & Gatto, M.A. (1979) Path dependent options: buy at the low, sell at the high. J. Fin. **34** 1111–1128.

Harper, J. (1994) Reducing parabolic partial differential equations to canonical form. Europ. J. Appl. Maths. **5** 159–164.

Harrison, J.M. & Kreps, D. (1979) Martingales and arbitrage in multiperiod securities markets. J. Econ. Theory **20** 381–408.

Harrison, J.M. & Pliska, S.R. (1981) Martingales and stochastic integrals in the theory of continuous trading. Stoch. Proc. Appl. **11** 215–260.

Hill, J.M. & Dewynne, J.N. (1990) *Heat Conduction*, CRC Press.

Ho, T. & Lee, S. (1986) Term structure movements and pricing interest rate contingent claims. J. Fin. **42** 1129–1142.

Hodges, S. & Neuberger, A. (1989) Optimal replication of contingent claims under transaction costs. Working paper, FORC, Warwick.

Hoggard, T., Whalley, A.E. & Wilmott, P. (1993) Hedging option portfolios in the presence of transaction costs. Adv. Fut. Opt. Res., forthcoming.

Hull, J. (1993) *Options, Futures and Other Derivative Securities*, second edition, Prentice-Hall.

Hull, J. & White, A. (1987) The pricing of options on assets with stochastic volatilities. J. Fin. **42** 281–300.

Hull, J. & White, A. (1990) Pricing interest rate derivative securities. Rev. Fin. Stud. **3** 573–92.

Ingersoll, J.E. (1987) *Theory of Financial Decision Making*, Rowman & Littlefield.

Jarrow, R.A. & Rudd, A. (1983) *Option Pricing*, Irwin.

Johnson, H.E. (1983) An analytical approximation to the American put price. J. Finan. Quant. An. **18** 141–148.

Johnson, L.W. & Riess, R.D. (1982) *Numerical Analysis* (2nd edition), Addison-Wesley.

Keener, J. (1988) *Principles of Applied Mathematics*, Addison-Wesley.

Kemna, A.G.Z. & Vorst, A.C.F. (1990) A pricing method for options based upon average asset values. J. Bank. Fin. March 113–129.

Kevorkian, J. (1990) *Partial Differential Equations: Analytical solution techniques*, Wadsworth and Brooks/Cole, Belmont, Calif.

Kinderlehrer, D. & Stampacchia, G. (1980) *An Introduction to Variational Inequalities and their Applications*, Academic Press.

Klemkosky, R.C. & Resnick, B.G. (1979) Put-call parity and market efficiency. J. Fin. **34** 1141–1155.

Klugman, R. (1992) Interest rate modelling, OCIAM working paper, Mathematical Institute, Oxford University.

Klugman, R. & Wilmott, P. (1993) A four parameter model for interest rates. OCIAM working paper, Mathematical Institute, Oxford University.

Leland, H.E. (1985) Option pricing and replication with transaction costs. J. Fin. **40** 1283–1301.

Leong, K. (1993) Estimates, guesstimates and rules of thumb, in *From Black-Scholes to Black Holes*, *Risk* magazine and Finex.

Levy, E. (1990) Asian arithmetic. *Risk* magazine May 7–8.

MacBeth, J.D. & Merville, L.J. (1979) An empirical examination of the Black-Scholes call option pricing model. J. Fin. **34** 1173–1186.

MacMillan, L.G. (1980) *Options as a Strategic Investment*, New York Inst. Fin., NY.

Mandelbrot, B. (1963) The variation of certain speculative prices. J. Bus. **36** 394–419.

Merton, R.C. (1973) Theory of rational option pricing. Bell J. Econ. Manag. Sci. **4** 141–83.

Merton, R.C. (1976) Option pricing when underlying stock returns are discontinuous. J. Finan. Econ. **3** 125–144.
Merton, R.C. (1990) *Continuous Time Finance*, Blackwell.
Øksendal, B. (1992) *Stochastic Differential Equations,* (third edition), Springer.
Pearson, N. & Sun, T-S. (1989) A test of the Cox, Ingersoll, Ross model of the term structure of interest rates using the method of moments. Sloan School of Management, MIT.
Richards, J.I. & Youn, H.K (1990) *Theory of Distributions*, Cambridge University Press.
Richtmyer, R.D. & Morton, K.W. (1967) *Difference Methods for Initial-value Problems*, Wiley Interscience.
Risk Magazine (1993) *From Black-Scholes to Black Holes*, *Risk* Magazine and Finex.
Roll, R. (1977) An analytical formula for unprotected American call options on stocks with known dividends. J. Finan. Econ. **5** 251–258.
Rubinstein, M. (1992) Exotic options. FORC Conf., Warwick.
Schuss, Z. (1980) *Theory and Applications of Stochastic Differential Equations*, Wiley.
Sharpe, W.F. (1985) *Investments*, Prentice-Hall
Smith, G.D. (1985) *Numerical Solution of Partial Differential Equations: Finite Difference Methods* (3rd edition), Oxford University Press.
Spiegel, M. (1980) *Probability and Statistics*, McGraw-Hill.
Stephenson, G. (1980) *Partial Differential Equations for Scientists and Engineers*, Longman.
Stoer, J. & Bulirsch, R. (1993) *Introduction to Numerical Analysis*, second edition, Springer.
Strang, G. (1986) *Introduction to Applied Mathematics*, Wellesley-Cambridge Press.
Stulz, R.M. (1982) Options on the minimum or maximum of two risky assets. J. Finan. Econ. **10** 161–185.
Vasicek, O.A. (1977) An equilibrium characterization of the term structure. J. Fin. Econ. **5** 177–188.
Whaley, R. (1981) On the valuation of American call options on stocks with known dividends. J. Finan. Econ. **9** 207–211.
Whalley, A.E. & Wilmott, P. (1993) A hedging strategy and option valuation model with transaction costs. OCIAM working paper, Mathematical Institute, Oxford University.
Whalley, A.E. & Wilmott, P. (1994a) An asymptotic analysis of the Davis, Panas and Zariphopoulou model for option pricing with transaction costs. OCIAM working paper, Mathematical Institute, Oxford University.
Whalley, A.E. & Wilmott, P. (1994b) Optional hedging with small but arbitrary transaction cost structure. OCIAM working paper, Mathematical Institute, Oxford University.
Williams, W.E. (1980) *Partial Differential Equations*, Oxford University Press.
Wilmott, P., Dewynne, J.N. & Howison, S.D. (1993) *Option Pricing: Mathematical Models and Computation.* Oxford Financial Press.

Index